Transportation
in the Future

STEM Road Map for Elementary School

Grade
3

Transportation in the Future

Grade 3

STEM Road Map for Elementary School

Edited by Carla C. Johnson, Janet B. Walton, and
Erin Peters-Burton

National Science Teachers Association

Arlington, Virginia

National Science Teachers Association

Claire Reinburg, Director
Rachel Ledbetter, Managing Editor
Deborah Siegel, Associate Editor
Amanda Van Beuren, Associate Editor
Donna Yudkin, Book Acquisitions Manager

ART AND DESIGN
Will Thomas Jr., Director, cover and
   interior design
Himabindu Bichali, Graphic Designer, interior
   design

PRINTING AND PRODUCTION
Catherine Lorrain, Director

NATIONAL SCIENCE TEACHERS ASSOCIATION
David L. Evans, Executive Director
David Beacom, Publisher

1840 Wilson Blvd., Arlington, VA 22201
www.nsta.org/store
For customer service inquiries, please call 800-277-5300.

NSTA is committed to publishing material that promotes the best in inquiry-based science education. However, conditions of actual use may vary, and the safety procedures and practices described in this book are intended to serve only as a guide. Additional precautionary measures may be required. NSTA and the authors do not warrant or represent that the procedures and practices in this book meet any safety code or standard of federal, state, or local regulations. NSTA and the authors disclaim any liability for personal injury or damage to property arising out of or relating to the use of this book, including any of the recommendations, instructions, or materials contained therein.

PERMISSIONS
Book purchasers may photocopy, print, or e-mail up to five copies of an NSTA book chapter for personal use only; this does not include display or promotional use. Elementary, middle, and high school teachers may reproduce forms, sample documents, and single NSTA book chapters needed for classroom or noncommercial, professional-development use only. E-book buyers may download files to multiple personal devices but are prohibited from posting the files to third-party servers or websites, or from passing files to non-buyers. For additional permission to photocopy or use material electronically from this NSTA Press book, please contact the Copyright Clearance Center (CCC) (*www.copyright.com*; 978-750-8400). Please access *www.nsta.org/permissions* for further information about NSTA's rights and permissions policies.

Library of Congress Cataloging-in-Publication Data
Names: Johnson, Carla C., 1969- editor. | Walton, Janet B., 1968- editor. | Peters-Burton, Erin E., editor.
Title: Transportation in the future, grade 3 : STEM road map for elementary school / edited by Carla C. Johnson, Janet B. Walton, and Erin Peters-Burton.
Description: Arlington, VA : National Science Teachers Association, [2017] | Series: STEM road map curriculum series | Includes bibliographical references and index.
Identifiers: LCCN 2017027984 (print) | LCCN 2017021515 (ebook) | ISBN 9781681403991 (print) | ISBN 9781681404004 (e-book)
Subjects: LCSH: Railroads--Study and teaching (Elementary)--United States. | Railroad travel--Study and teaching (Elementary)--United States. | Transportation--Study and teaching (Elementary)--United States | Transportation--United States--Forecasting. | Third grade (Education)
Classification: LCC TF173 .T73 2017 (ebook) | LCC TF173 (print) | DDC 372.35/8--dc23
LC record available at *https://lccn.loc.gov/2017027984*

The *Next Generation Science Standards* ("NGSS") were developed by twenty-six states, in collaboration with the National Research Council, the National Science Teachers Association and the American Association for the Advancement of Science in a process managed by Achieve, Inc. For more information go to *www.nextgenscience.org*.

CONTENTS

CONTENTS

ABOUT THE EDITORS AND AUTHORS

Dr. Carla C. Johnson is the associate dean for research, engagement, and global partnerships and a professor of science education at Purdue University's College of Education in West Lafayette, Indiana. Dr. Johnson serves as the director of research and evaluation for the Department of Defense–funded Army Educational Outreach Program (AEOP), a global portfolio of STEM education programs, competitions, and apprenticeships. She has been a leader in STEM education for the past decade, serving as the director of STEM Centers, editor of the *School Science and Mathematics* journal, and lead researcher for the evaluation of Tennessee's Race to the Top–funded STEM portfolio. Dr. Johnson has published over 100 articles, books, book chapters, and curriculum books focused on STEM education. She is a former science and social studies teacher and was the recipient of the 2013 Outstanding Science Teacher Educator of the Year award from the Association for Science Teacher Education (ASTE), the 2012 Award for Excellence in Integrating Science and Mathematics from the School Science and Mathematics Association (SSMA), the 2014 award for best paper on Implications of Research for Educational Practice from ASTE, and the 2006 Outstanding Early Career Scholar Award from SSMA. Her research focuses on STEM education policy implementation, effective science teaching, and integrated STEM approaches.

Dr. Janet B. Walton is the research assistant professor and the assistant director of evaluation for AEOP at Purdue University's College of Education. Formerly the STEM workforce program manager for Virginia's Region 2000 and founding director of the Future Focus Foundation, a nonprofit organization dedicated to enhancing the quality of STEM education in the region, she merges her economic development and education backgrounds to develop K–12 curricular materials that integrate real-life issues with sound cross-curricular content. Her research focuses on collaboration between schools and community stakeholders for STEM education and problem- and project-based learning pedagogies. With this research agenda, she works to forge productive relationships between K–12 schools and local business and community stakeholders to bring contextual STEM experiences into the classroom and provide students and educators with innovative resources and curricular materials.

Dr. Erin Peters-Burton is the Donna R. and David E. Sterling endowed professor in science education at George Mason University in Fairfax, Virginia. She uses her experiences from 15 years as an engineer and secondary science, engineering, and mathematics teacher to develop research projects that directly inform classroom practice in science and engineering. Her research agenda is based on the idea that all students should build self-awareness of how they learn science and engineering. She works to help students see themselves as "science-minded" and help teachers create classrooms that support student skills to develop scientific knowledge. To accomplish this, she pursues research projects that investigate ways that students and teachers can use self-regulated learning theory in science and engineering, as well as how inclusive STEM schools can help students succeed. During her tenure as a secondary teacher, she had a National Board Certification in Early Adolescent Science and was an Albert Einstein Distinguished Educator Fellow for NASA. As a researcher, Dr. Peters-Burton has published over 100 articles, books, book chapters, and curriculum books focused on STEM education and educational psychology. She received the Outstanding Science Teacher Educator of the Year award from ASTE in 2016 and a Teacher of Distinction Award and a Scholarly Achievement Award from George Mason University in 2012, and in 2010 she was named University Science Educator of the Year by the Virginia Association of Science Teachers.

Andrea R. Milner is the vice president and dean of academic affairs and an associate professor in the Teacher Education Department at Adrian College in Adrian, Michigan. A former early childhood and elementary teacher, Dr. Milner researches the effects constructivist classroom contextual factors have on student motivation and learning strategy use.

Tamara J. Moore is an associate professor of engineering education in the College of Engineering at Purdue University. Dr. Moore's research focuses on defining STEM integration through the use of engineering as the connection and investigating its power for student learning.

Toni A. Sondergeld is an associate professor of assessment, research, and statistics in the School of Education at Drexel University in Philadelphia. Dr. Sondergeld's research concentrates on assessment and evaluation in education, with a focus on K–12 STEM.

Sandy Watkins is principal-in-residence for the Tennessee STEM Innovation Network. Ms. Watkins has been a science educator for kindergarten through postgraduate-level classes and has served as a STEM coordinator and consultant. She was also the founding principal of the Innovation Academy of Northeas Tennessee, which is a STEM school.

ACKNOWLEDGMENTS

This module was developed as a part of the STEM Road Map project (Carla C. Johnson, principal investigator). The Purdue University College of Education, General Motors, and other sources provided funding for this project.

See *www.routledge.com/products/9781138804234* for more information about *STEM Road Map: A framework for integrated STEM education*.

PART 1

THE STEM ROAD MAP

BACKGROUND, THEORY, AND PRACTICE

OVERVIEW OF THE *STEM ROAD MAP CURRICULUM SERIES*

Carla C. Johnson, Erin E. Peters-Burton, and Tamara J. Moore

The *STEM Road Map Curriculum Series* was conceptualized and developed by a team of STEM educators from across the United States in response to a growing need to infuse real-world learning contexts, delivered through authentic problem-solving pedagogy, into K–12 classrooms. The curriculum series is grounded in integrated STEM, which focuses on the integration of the STEM disciplines—science, technology, engineering, and mathematics—delivered across content areas, incorporating the Framework for 21st Century Learning along with grade-level-appropriate academic standards.

The curriculum series begins in kindergarten, with a five-week instructional sequence that introduces students to the STEM themes and gives them grade-level-appropriate topics and real-world challenges or problems to solve. The series uses project- and problem-based learning, presenting students with the problem or challenge during the first lesson, and then teaching them science, social studies, English language arts, mathematics, and other content, as they apply what they learn to the challenge or problem at hand.

Authentic assessment and differentiation are embedded throughout the modules. Each *STEM Road Map Curriculum Series* module has a lead discipline, which may be science, social studies, English language arts, or mathematics. All disciplines are integrated into each module, along with ties to engineering. Another key component is the use of STEM Research Notebooks to allow students to track their own learning progress. The modules are designed with a scaffolded approach, with increasingly complex concepts and skills introduced as students progress through grade levels.

The developers of this work view the curriculum as a resource that is intended to be used either as a whole or in part to meet the needs of districts, schools, and teachers who are implementing an integrated STEM approach. A variety of implementation formats are possible, from using one stand-alone module at a given grade level to using all five modules to provide 25 weeks of instruction. Also, within each grade band (K–2, 3–5, 6–8, 9–12), the modules can be sequenced in various ways to suit specific needs.

STANDARDS-BASED APPROACH

The *STEM Road Map Curriculum Series* is anchored in the *Next Generation Science Standards (NGSS)*, the *Common Core State Standards for Mathematics (CCSS Mathematics)*, the *Common Core State Standards for English Language Arts (CCSS ELA)*, and the Framework for 21st Century Learning. Each module includes a detailed curriculum map that incorporates the associated standards from the particular area correlated to lesson plans. The STEM Road Map has very clear and strong connections to these academic standards, and each of the grade-level topics was derived from the mapping of the standards to ensure alignment among topics, challenges or problems, and the required academic standards for students. Therefore, the curriculum series takes a standards-based approach and is designed to provide authentic contexts for application of required knowledge and skills.

THEMES IN THE *STEM ROAD MAP CURRICULUM SERIES*

The K–12 STEM Road Map is organized around five real-world STEM themes that were generated through an examination of the big ideas and challenges for society included in STEM standards and those that are persistent dilemmas for current and future generations:

- Cause and Effect
- Innovation and Progress
- The Represented World
- Sustainable Systems
- Optimizing the Human Experience

These themes are designed as springboards for launching students into an exploration of real-world learning situated within big ideas. Most important, the five STEM Road Map themes serve as a framework for scaffolding STEM learning across the K–12 continuum.

The themes are distributed across the STEM disciplines so that they represent the big ideas in science (Cause and Effect; Sustainable Systems), technology (Innovation and Progress; Optimizing the Human Experience), engineering (Innovation and Progress; Sustainable Systems; Optimizing the Human Experience), and mathematics (The Represented World), as well as concepts and challenges in social studies and 21st century skills that are also excellent contexts for learning in English language arts. The process of developing themes began with the clustering of the *NGSS* performance expectations and the National Academy of Engineering's grand challenges for engineering, which led to the development of the challenge in each module and connections of the module activities to the *CCSS Mathematics* and *CCSS ELA* standards. We performed these

mapping processes with large teams of experts and found that these five themes provided breadth, depth, and coherence to frame a quality STEM learning experience from kindergarten through twelfth grade.

Cause and Effect

The concept of cause and effect is a powerful and pervasive notion in the STEM fields. It is the foundation of understanding how and why things happen as they do. Humans spend considerable effort and resources trying to understand the causes and effects of natural and designed phenomena to gain better control over events and the environment and to be prepared to react appropriately. Equipped with the knowledge of a specific cause-and-effect relationship, we can lead better lives or contribute to the community by altering the cause, leading to a different effect. For example, if a person recognizes that irresponsible energy consumption leads to global climate change, that person can act to remedy his or her contribution to the situation. Although cause and effect is a core idea in the STEM fields, it can actually be very difficult to determine. Students should be capable of understanding not only when evidence points to cause and effect but also when evidence points to relationships but not direct causality. The major goal of education is to foster students to be empowered, analytic thinkers, capable of thinking through complex processes to make important decisions. Understanding causality, as well as when it cannot be determined, will help students become better consumers, global citizens, and community members.

Innovation and Progress

One of the most important factors in determining whether humans will have a positive future is innovation. Innovation is the driving force behind progress, which helps create possibilities that did not exist before. Innovation and progress are creative entities, but in the STEM fields, they are anchored by evidence and logic, and they use established concepts to move the STEM fields forward. In creating something new, students must consider what is already known in the STEM fields and apply this knowledge appropriately. When we innovate, we create value that was not there previously and create new conditions and possibilities for even more innovations. Students should consider how their innovations might affect progress and use their STEM thinking to change current human burdens to benefits. For example, if we develop more efficient cars that use byproducts from another manufacturing industry, such as food processing, then we have used waste productively and reduced the need for the waste to be hauled away, an indirect benefit of the innovation.

The Represented World

When we communicate about the world we live in, how the world works, and how we can meet the needs of humans, sometimes we can use the actual phenomena to explain a concept. Sometimes, however, the concept is too big, too slow, too small, too fast, or too complex for us to explain using the actual phenomena, and we must use a representation or a model to help communicate the important features. We need representations and models such as graphs, tables, mathematical expressions, and diagrams because it makes our thinking visible. For example, when examining geologic time, we cannot actually observe the passage of such large chunks of time, so we create a timeline or a model that uses a proportional scale to visually illustrate how much time has passed for different eras. Another example may be something too complex for students at a particular grade level, such as explaining the *p* subshell orbitals of electrons to fifth graders. Instead, we use the Bohr model, which more closely represents the orbiting of planets and is accessible to fifth graders.

When we create models, they are helpful because they point out the most important features of a phenomenon. We also create representations of the world with mathematical functions, which help us change parameters to suit the situation. Creating representations of a phenomenon engages students because they are able to identify the important features of that phenomenon and communicate them directly. But because models are estimates of a phenomenon, they leave out some of the details, so it is important for students to evaluate their usefulness as well as their shortcomings.

Sustainable Systems

From an engineering perspective, the term *system* refers to the use of "concepts of component need, component interaction, systems interaction, and feedback. The interaction of subcomponents to produce a functional system is a common lens used by all engineering disciplines for understanding, analysis, and design." (Koehler et al. 2013, p. 8). Systems can be either open (e.g., an ecosystem) or closed (e.g., a car battery). Ideally, a system should be sustainable, able to maintain equilibrium without much energy from outside the structure. Looking at a garden, we see flowers blooming, weeds sprouting, insects buzzing, and various forms of life living within its boundaries. This is an example of an ecosystem, a collection of living organisms that survive together, functioning as a system. The interaction of the organisms within the system and the influences of the environment (e.g., water, sunlight) can maintain the system for a period of time, thus demonstrating its ability to endure. Sustainability is a desirable feature of a system because it allows for existence of the entity in the long term.

In the STEM Road Map project, we identified different standards that we consider to be oriented toward systems that students should know and understand in the K–12 setting. These include ecosystems, the rock cycle, Earth processes (such as erosion, tectonics, ocean currents, weather phenomena), Earth-Sun-Moon cycles, heat transfer, and the interaction among the geosphere, biosphere, hydrosphere, and atmosphere. Students and teachers should understand that we live in a world of systems that are not independent of each other, but rather are intrinsically linked such that a disruption in one part of a system will have reverberating effects on other parts of the system.

Optimizing the Human Experience

Science, technology, engineering, and mathematics as disciplines have the capacity to continuously improve the ways humans live, interact, and find meaning in the world, thus working to optimize the human experience. This idea has two components: being more suited to our environment and being more fully human. For example, the progression of STEM ideas can help humans create solutions to complex problems, such as improving ways to access water sources, designing energy sources with minimal impact on our environment, developing new ways of communication and expression, and building efficient shelters. STEM ideas can also provide access to the secrets and wonders of nature. Learning in STEM requires students to think logically and systematically, which is a way of knowing the world that is markedly different from knowing the world as an artist. When students can employ various ways of knowing and understand when it is appropriate to use a different way of knowing or integrate ways of knowing, they are fully experiencing the best of what it is to be human. The problem-based learning scenarios provided in the STEM Road Map help students develop ways of thinking like STEM professionals as they ask questions and design solutions. They learn to optimize the human experience by innovating improvements in the designed world in which they live.

THE NEED FOR AN INTEGRATED STEM APPROACH

At a basic level, STEM stands for science, technology, engineering, and mathematics. Over the past decade, however, STEM has evolved to have a much broader scope and implications. Now, educators and policymakers refer to STEM as not only a concentrated area for investing in the future of the United States and other nations but also as a domain and mechanism for educational reform.

The good intentions of the recent decade-plus of focus on accountability and increased testing has resulted in significant decreases not only in instructional time for teaching science and social studies but also in the flexibility of teachers to promote authentic problem solving focused classroom environments. The shift has had a detrimental impact on student acquisition of vitally important skills, which many refer to as 21st century skills, and often the ability of students to "think." Further, schooling has become increasingly siloed into compartments of mathematics, science, English language arts, and social studies, lacking any of the connections that are overwhelmingly present in the real world

around children. Students have experienced school as content provided in boxes that must be memorized, devoid of any real-world context, and often have little understanding of why they are learning these things.

STEM-focused projects, curriculum, activities, and schools have emerged as a means to address these challenges. However, most of these efforts have continued to focus on the individual STEM disciplines (predominantly science and engineering) through more STEM classes and after-school programs in a "STEM enhanced" approach (Breiner et al. 2012). But in traditional and STEM enhanced approaches, there is little to no focus on other disciplines that are integral to the context of STEM in the real world. Integrated STEM education, on the other hand, infuses the learning of important STEM content and concepts with a much-needed emphasis on 21st century skills and a problem- and project-based pedagogy that more closely mirrors the real-world setting for society's challenges. It incorporates social studies, English language arts, and the arts as pivotal and necessary (Johnson 2013; Rennie, Venville, and Wallace 2012; Roehrig et al. 2012).

FRAMEWORK FOR STEM INTEGRATION IN THE CLASSROOM

The *STEM Road Map Curriculum Series* is grounded in the Framework for STEM Integration in the Classroom as conceptualized by Moore, Guzey, and Brown (2014) and Moore et al. (2014). The framework has six elements, described in the context of how they are used in the *STEM Road Map Curriculum Series* as follows:

1. The STEM Road Map contexts are meaningful to students and provide motivation to engage with the content. Together, these allow students to have different ways to enter into the challenge.

2. The STEM Road Map modules include engineering design that allows students to design technologies (i.e., products that are part of the designed world) for a compelling purpose.

3. The STEM Road Map modules provide students with the opportunities to learn from failure and redesign based on the lessons learned.

4. The STEM Road Map modules include standards-based disciplinary content as the learning objectives.

5. The STEM Road Map modules include student-centered pedagogies that allow students to grapple with the content, tie their ideas to the context, and learn to think for themselves as they deepen their conceptual knowledge.

6. The STEM Road Map modules emphasize 21st century skills and, in particular, highlight communication and teamwork.

All of the STEM Road Map modules incorporate these six elements; however, the level of emphasis on each of these elements varies based on the challenge or problem in each module.

THE NEED FOR THE *STEM ROAD MAP CURRICULUM SERIES*

As focus is increasing on integrated STEM, and additional schools and programs decide to move their curriculum and instruction in this direction, there is a need for quality, research-based curriculum designed with integrated STEM at the core. Several good resources are available to help teachers infuse engineering or more STEM enhanced approaches, but no curriculum exists that spans K–12 with an integrated STEM focus. The next chapter provides detailed information about the specific pedagogy, instructional strategies, and learning theory on which the *STEM Road Map Curriculum Series* is grounded.

REFERENCES

Breiner, J., M. Harkness, C. C. Johnson, and C. Koehler. 2012. What is STEM? A discussion about conceptions of STEM in education and partnerships. *School Science and Mathematics* 112 (1): 3–11.

Johnson, C. C. 2013. Conceptualizing integrated STEM education: Editorial. *School Science and Mathematics* 113 (8): 367–368.

Koehler, C. M., M. A. Bloom, and I. C. Binns. 2013. Lights, camera, action: Developing a methodology to document mainstream films' portrayal of nature of science and scientific inquiry. *Electronic Journal of Science Education* 17 (2).

Krajcik, J., and P. Blumenfeld. 2006. Project-based learning. In *The Cambridge handbook of the learning sciences,* ed. R. Keith Sawyer, 317–334. New York: Cambridge University Press.

Lambros, A. 2004. *Problem-based learning in middle and high school classrooms: A teacher's guide to implementation.* Thousand Oaks, CA: Corwin Press.

Moore, T. J., S. S. Guzey, and A. Brown. 2014. Greenhouse design to increase habitable land: An engineering unit. *Science Scope* 51–57.

Moore, T. J., M. S. Stohlmann, H.-H. Wang, K. M. Tank, A. W. Glancy, and G. H. Roehrig. 2014. Implementation and integration of engineering in K–12 STEM education. In *Engineering in pre-college settings: Synthesizing research, policy, and practices,* ed. S. Purzer, J. Strobel, and M. Cardella, 35–60. West Lafayette, IN: Purdue Press.

Rennie, L., G. Venville, and J. Wallace. 2012. *Integrating science, technology, engineering, and mathematics: Issues, reflections, and ways forward.* New York: Routledge.

Roehrig, G. H., T. J. Moore, H. H. Wang, and M. S. Park. 2012. Is adding the *E* enough? Investigating the impact of K–12 engineering standards on the implementation of STEM integration. *School Science and Mathematics* 112 (1): 31–44.

STRATEGIES USED IN THE *STEM ROAD MAP CURRICULUM SERIES*

Erin E. Peters-Burton, Carla C. Johnson, Toni A. Sondergeld, and Tamara J. Moore

The *STEM Road Map Curriculum Series* uses what has been identified through research as best-practice pedagogy, including embedded formative assessment strategies throughout each module. This chapter briefly describes the key strategies that are employed in the series.

PROJECT- AND PROBLEM-BASED LEARNING

Each module in the *STEM Road Map Curriculum Series* uses either project-based learning or problem-based learning to drive the instruction. Project-based learning begins with a driving question to guide student teams in addressing a contextualized local or community problem or issue. The outcome of project-based instruction is a product that is conceptualized, designed, and tested through a series of scaffolded learning experiences (Blumenfeld et al. 1991; Krajcik and Blumenfeld 2006). Problem-based learning is often grounded in a fictitious scenario, challenge, or problem (Barell 2006; Lambros 2004). On the first day of instruction within the unit, student teams are provided with the context of the problem. Teams work through a series of activities and use open-ended research to develop their potential solution to the problem or challenge, which may not be a tangible product (Johnson 2003).

ENGINEERING DESIGN PROCESS

The *STEM Road Map Curriculum Series* uses engineering design as a way to facilitate integrated STEM within the modules. The engineering design process (EDP) is depicted in Figure 2.1 (p. 10). It highlights two major aspects of engineering design—problem scoping and solution generation—and six specific components of working toward a design: define the problem, learn about the problem, plan a solution, try the solution, test the solution, decide whether the solution is good enough. It also shows that communication

Figure 2.1. Engineering Design Process

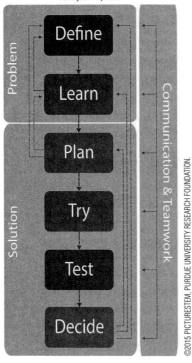

and teamwork are involved throughout the entire process. As the arrows in the figure indicate, the order in which the components of engineering design are addressed depends on what becomes needed as designers progress through the EDP. Designers must communicate and work in teams throughout the process. The EDP is iterative, meaning that components of the process can be repeated as needed until the design is good enough to present to the client as a potential solution to the problem.

Problem scoping is the process of gathering and analyzing information to deeply understand the engineering design problem. It includes defining the problem and learning about the problem. Defining the problem includes identifying the problem, the client, and the end user of the design. The client is the person (or people) who hired the designers to do the work, and the end user is the person (or people) who will use the final design. The designers must also identify the criteria and the constraints of the problem. The criteria are the things the client wants from the solution, and the constraints are the things that limit the possible solutions. The designers must spend significant time learning about the problem, which can include activities such as the following:

- Reading informational texts and researching about relevant concepts or contexts

- Identifying and learning about needed mathematical and scientific skills, knowledge, and tools

- Learning about things done previously to solve similar problems

- Experimenting with possible materials that could be used in the design

Problem scoping also allows designers to consider how to measure the success of the design in addressing specific criteria and staying within the constraints over multiple iterations of solution generation.

Solution generation includes planning a solution, trying the solution, testing the solution, and deciding whether the solution is good enough. Planning the solution includes generating many design ideas that both address the criteria and meet the constraints. Here the designers must consider what was learned about the problem during problem scoping. Design plans include clear communication of design ideas through media such as notebooks, blueprints, schematics, or storyboards. They also include details about the design, such as measurements, materials, colors, costs of materials, instructions for how things fit together, and sets of directions. Making the decision about which design idea to move forward involves considering the trade-offs of each design idea.

Once a clear design plan is in place, the designers must try the solution. Trying the solution includes developing a prototype (a testable model) based on the plan generated. The prototype might be something physical or a process to accomplish a goal. This component of design requires that the designers consider the risk involved in implementing the design. The prototype developed must be tested. Testing the solution includes conducting fair tests that verify whether the plan is a solution that is good enough to meet the client and end user needs and wants. Data need to be collected about the results of the tests of the prototype, and these data should be used to make evidence-based decisions regarding the design choices made in the plan. Here, the designers must again consider the criteria and constraints for the problem.

Using the data gathered from the testing, the designers must decide whether the solution is good enough to meet the client and end user needs and wants by assessment based on the criteria and constraints. Here, the designers must justify or reject design decisions based on the background research gathered while learning about the problem and on the evidence gathered during the testing of the solution. The designers must now decide whether to present the current solution to the client as a possibility or to do more iterations of design on the solution. If they decide that improvements need to be made to the solution, the designers must decide if there is more that needs to be understood about the problem, client, or end user; if another design idea should be tried; or if more planning needs to be conducted on the same design. One way or another, more work needs to be done.

Throughout the process of designing a solution to meet a client's needs and wants, designers work in teams and must communicate to each other, the client, and likely the end user. Teamwork is important in engineering design because multiple perspectives and differing skills and knowledge are valuable when working to solve problems. Communication is key to the success of the designed solution. Designers must communicate their ideas clearly using many different representations, such as text in an engineering notebook, diagrams, flowcharts, technical briefs, or memos to the client.

LEARNING CYCLE

The same format for the learning cycle is used in all grade levels throughout the STEM Road Map, so that students engage in a variety of activities to learn about phenomena in the modules thoroughly and have consistent experiences in the problem- and project-based learning modules. Expectations for learning by younger students are not as high as for older students, but the format of the progression of learning is the same. Students who have learned with curriculum from the STEM Road Map in early grades know

what to expect in later grades. The learning cycle consists of five parts—Introductory Activity/Engagement, Activity/Exploration, Explanation, Elaboration/Application of Knowledge, and Evaluation/Assessment—and is based on the empirically tested 5E model from BSCS (Bybee et al. 2006).

In the Introductory Activity/Engagement phase, teachers introduce the module challenge and use a unique approach designed to pique students' curiosity. This phase gets students to start thinking about what they already know about the topic and begin wondering about key ideas. The Introductory Activity/Engagement phase positions students to be confident about what they are about to learn, because they have prior knowledge, and clues them into what they don't yet know.

In the Activity/Exploration phase, the teacher sets up activities in which students experience a deeper look at the topics that were introduced earlier. Students engage in the activities and generate new questions or consider possibilities using preliminary investigations. Students work independently, in small groups, and in whole-group settings to conduct investigations, resulting in common experiences about the topic and skills involved in the real-world activities. Teachers can assess students' development of concepts and skills based on the common experiences during this phase.

During the Explanation phase, teachers direct students' attention to concepts they need to understand and skills they need to possess to accomplish the challenge. Students participate in activities to demonstrate their knowledge and skills to this point, and teachers can pinpoint gaps in student knowledge during this phase.

In the Elaboration/Application phase, teachers present students with activities that engage in higher-order thinking to create depth and breadth of student knowledge, while connecting ideas across topics within and across STEM. Students apply what they have learned thus far in the module to a new context or elaborate on what they have learned about the topic to a deeper level of detail.

In the last phase, Assessment, teachers give students summative feedback on their knowledge and skills as demonstrated through the challenge. This is not the only point of assessment (as discussed in the section on Embedded Formative Assessments below), but it is an assessment of the culmination of the knowledge and skills for the module. Students demonstrate their cognitive growth at this point and reflect on how far they have come since the beginning of the module. The challenges are designed to be multidimensional in the ways students must collaborate and communicate their new knowledge.

STEM RESEARCH NOTEBOOK

One of the main components of the *STEM Road Map Curriculum Series* is the STEM Research Notebook, a place for students to capture their ideas, questions, observations, reflections, evidence of progress, and other items associated with their daily work. At the beginning of each module, the teacher walks students through the setup of the STEM

Research Notebook, which could be a three ring-binder, composition book, or spiral notebook. You may wish to have students create divided sections so that they can easily access work from various disciplines during the module. Electronic notebooks kept on student devices are also acceptable and encouraged. Students will develop their own table of contents and create chapters in the notebook for each module.

Each lesson in the *STEM Road Map Curriculum Series* includes one or more prompts that are designed for inclusion in the STEM Research Notebook and appear as questions or statements that the teacher assigns to students. These prompts require students to apply what they have learned across the lesson to solve the big problem or challenge for that module. Each lesson is designed to meaningfully refer students to the larger problem or challenge they have been assigned to solve with their teams. The STEM Research Notebook is designed to be a key formative assessment tool, as students' daily entries provide evidence of what they are learning. The notebook can be used as a mechanism for dialogue between the teacher and students, as well as for peer and self-evaluation.

The use of the STEM Research Notebook is designed to scaffold student notebooking skills across the grade bands in the *STEM Road Map Curriculum Series*. In the early grades, children learn how to organize their daily work in the notebook as a way to collect their products for future reference. In elementary school, students structure their notebooks to integrate background research along with their daily work and lesson prompts. In the upper grades (middle and high school), students expand their use of research and data gathering through team discussions to more closely mirror the work of STEM experts in the real world.

THE ROLE OF ASSESSMENT IN THE *STEM ROAD MAP CURRICULUM SERIES*

Starting in the middle years and continuing into secondary education, the word *assessment* typically brings grades to mind. These grades may take the form of a letter or a percentage, but they typically are used as a representation of a student's content mastery. If well thought out and implemented, however, classroom assessment can offer teachers, parents, and students valuable information about student learning and misconceptions that does not necessarily come in the form of a grade (Popham 2013).

The *STEM Road Map Curriculum Series* provides a set of assessments for each module. Teachers are encouraged to use assessment information for more than just assigning grades to students. Instead, assessments of activities requiring students to actively engage in their learning, such as student journaling in STEM Research Notebooks, collaborative presentations, and constructing graphic organizers, should be used to move student learning forward. Whereas other curriculum series with assessments may include objective-type (multiple-choice or matching) tests, quizzes, or worksheets, we have intentionally avoided these forms of assessments to better align assessment strategies with

teacher instruction and student learning techniques. Since the focus of this book is on project- or problem-based STEM curriculum and instruction that focuses on higher-level thinking skills, appropriate and authentic performance assessments were developed to elicit the most reliable and valid indication of growth in student abilities (Brookhart and Nitko 2008).

Comprehensive Assessment System

Assessment throughout all STEM Road Map curriculum modules acts as a comprehensive system in which both formative and summative assessments work together to provide teachers with quality information on student learning. Formative assessment occurs when the teacher finds out formally or informally what a student knows about a smaller, defined concept or skill and provides timely feedback to the student about his or her level of proficiency. Summative assessments occur when students have performed all activities in the module and are given a cumulative performance evaluation in which they demonstrate their growth in learning.

A comprehensive assessment system can be thought of as akin to a sporting event. Formative assessments are the practices: It is important to accomplish them consistently, they provide feedback to help students improve their learning, and making mistakes can be worthwhile if students are given an opportunity to learn from them. Summative assessments are the competitions: Students need to be prepared to perform at the best of their ability. Without multiple opportunities to practice skills along the way through formative assessments, students will not have the best chance of demonstrating growth in abilities through summative assessments (Black and Wiliam 1998).

Embedded Formative Assessments

Formative assessments in this module serve two main purposes: to provide feedback to students about their learning and to provide important information for the teacher to inform immediate instructional needs. Providing feedback to students is particularly important when conducting problem- or project-based learning because students take on much of the responsibility for learning, and teachers must facilitate student learning in an informed way. For example, if students are required to conduct research for the Activity/Exploration phase but are not familiar with what constitutes a reliable resource, they may develop misconceptions based on poor information. When a teacher monitors this learning through formative assessments and provides specific feedback related to the instructional goals, students are less likely to develop incomplete or incorrect conceptions in their independent investigations. By using formative assessment to detect problems in student learning and then acting on this information, teachers help move student learning forward through these teachable moments.

Formative assessments come in a variety of formats. They can be informal, such as asking students probing questions related to student knowledge or tasks or simply observing students engaged in an activity to gather information about student skills. Formative assessments can also be formal, such as a written quiz or a laboratory practical. Regardless of the type, three key steps must be completed when using formative assessments (Sondergeld, Bell, and Leusner 2010). First, the assessment is delivered to students so that teachers can collect data. Next, teachers analyze the data (student responses) to determine student strengths and areas that need additional support. Finally, teachers use the results from information collected to modify lessons and create learning environments that reinforce weak points in student learning. If student learning information is not used to modify instruction, the assessment cannot be considered formative in nature.

Formative assessments can be about content, science process skills, or even learning skills. When a formative assessment focuses on content, it assesses student knowledge about the disciplinary core ideas from the *Next Generation Science Standards* (*NGSS*) or content objectives from Common Core mathematics or English language arts. Content-focused formative assessments ask students questions about declarative knowledge regarding the concepts they have been learning. Process skills formative assessments examine the extent to which a student can perform science and engineering practices from the *NGSS* or process objectives from *CCSS Mathematics* or *CCSS ELA*, such as constructing an argument. Learning skills can also be assessed formatively by asking students to reflect on the ways they learn best during a module and identify ways they could have learned more.

Assessment Maps

Assessment maps or blueprints can be used to ensure alignment between classroom instruction and assessment. If what students are learning in the classroom is not the same as the content on which they are assessed, the resultant judgment made on student learning will be invalid (Brookhart and Nitko 2008). Therefore, the issue of instruction and assessment alignment is critical. The assessment map for this book (found in Chapter 3) indicates by lesson whether the assessment should be completed as a group or on an individual basis, identifies the assessment as formative or summative in nature, and aligns the assessment with its corresponding learning objectives.

Note that the module includes far more formative assessments than summative assessments. This is done intentionally to provide students with multiple opportunities to practice their learning of new skills before completing a summative assessment. Note also that formative assessments are used to collect information on only one or two learning objectives at a time so that potential relearning or instructional modifications can focus on smaller and more manageable chunks of information. Conversely, summative assessments in the module cover many more learning objectives, as they are traditionally used as final

markers of student learning. This is not to say that information collected from summative assessments cannot or should not be used formatively. If teachers find that gaps in student learning persist after a summative assessment is completed, it is important to revisit these existing misconceptions or areas of weakness before moving on (Black et al. 2003).

SELF-REGULATED LEARNING THEORY IN THE STEM ROAD MAP MODULES

Many learning theories are compatible with the STEM Road Map modules, such as constructivism, situated cognition, and meaningful learning. However, we feel that the self-regulated learning theory (SRL) aligns most appropriately (Zimmerman 2000). SRL requires students to understand that thinking needs to be motivated and managed (Ritchhart, Church, and Morrison 2011). The STEM Road Map modules are student centered and are designed to provide students with choices, concrete hands-on experiences, and opportunities to see and make connections, especially across subjects (Eliason and Jenkins 2012; NAEYC 2016). Additionally, SRL is compatible with the modules because it fosters a learning environment that supports students' motivation, enables students to become aware of their own learning strategies, and requires reflection on learning while experiencing the module (Peters and Kitsantas 2010).

Figure 2.2. SRL Theory

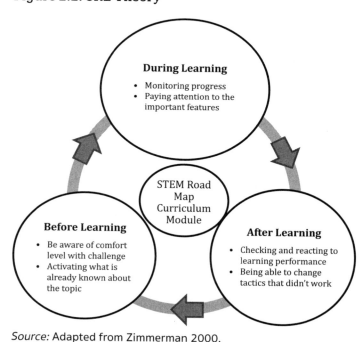

Source: Adapted from Zimmerman 2000.

The theory behind SRL (see Figure 2.2) explains the different processes that students engage in before, during, and after a learning task. Because SRL is a cyclical learning process, the accomplishment of one cycle develops strategies for the next learning cycle. This cyclic way of learning aligns with the various sections in the STEM Road Map lesson plans on Introductory Activity/ Engagement, Activity/Exploration, Explanation, Elaboration/Application of Knowledge, and Evaluation/Assessment. Since the students engaged in a module take on much of the responsibility for learning, this theory also provides guidance for teachers to keep students on the right track.

The remainder of this section explains how SRL theory is embedded within the five sections of each module and points out ways to support students in becoming independent learners of STEM while productively functioning in collaborative teams.

Before Learning: Setting the Stage

Before attempting a learning task such as the STEM Road Map modules, teachers should develop an understanding of their students' level of comfort with the process of accomplishing the learning and determine what they already know about the topic. When students are comfortable with attempting a learning task, they tend to take more risks in learning and as a result achieve deeper learning (Bandura 1986).

The STEM Road Map curriculum modules are designed to foster excitement from the very beginning. Each module has an Introductory Activity/Engagement section that introduces the overall topic from a unique and exciting perspective, engaging the students to learn more so that they can accomplish the challenge. The Introductory Activity also has a design component that helps teachers assess what students already know about the topic of the module. In addition to the deliberate designs in the lesson plans to support SRL, teachers can support a high level of student comfort with the learning challenge by finding out if students have ever accomplished the same kind of task and, if so, asking them to share what worked well for them.

During Learning: Staying the Course

Some students fear inquiry learning because they aren't sure what to do to be successful (Peters 2010). However, the STEM Road Map curriculum modules are embedded with tools to help students pay attention to knowledge and skills that are important for the learning task and to check student understanding along the way. One of the most important processes for learning is the ability for learners to monitor their own progress while performing a learning task (Peters 2012). The modules allow students to monitor their progress with tools such as the STEM Research Notebooks, in which they record what they know and can check whether they have acquired a complete set of knowledge and skills. The STEM Road Map modules support inquiry strategies that include previewing, questioning, predicting, clarifying, observing, discussing, and journaling (Morrison and Milner 2014). Through the use of technology throughout the modules, inquiry is supported by providing students access to resources and data while enabling them to process information, report the findings, collaborate, and develop 21st century skills.

It is important for teachers to encourage students to have an open mind about alternative solutions and procedures (Milner and Sondergeld 2015) when working through the STEM Road Map curriculum modules. Novice learners can have difficulty knowing what to pay attention to and tend to treat each possible avenue for information as equal (Benner 1984). Teachers are the mentors in a classroom and can point out ways

for students to approach learning during the Activity/Exploration, Explanation, and Elaboration/Application of Knowledge portions of the lesson plans to ensure that students pay attention to the important concepts and skills throughout the module. For example, if a student is to demonstrate conceptual awareness of motion when working on roller coaster research, but the student has misconceptions about motion, the teacher can step in and redirect student learning.

After Learning: Knowing What Works

The classroom is a busy place, and it may often seem that there is no time for self-reflection on learning. Although skipping this reflective process may save time in the short term, it reduces the ability to take into account things that worked well and things that didn't so that teaching the module may be improved next time. In the long run, SRL skills are critical for students to become independent learners who can adapt to new situations. By investing the time it takes to teach students SRL skills, teachers can save time later, because students will be able to apply methods and approaches for learning that they have found effective to new situations. In the Assessment portion of the STEM Road Map curriculum modules, as well as in the formative assessments throughout the modules, two processes in the after-learning phase are supported: evaluating one's own performance and accounting for ways to adapt tactics that didn't work well. Students have many opportunities to self-assess in formative assessments, both in groups and individually, using the rubrics provided in the modules.

The designs of the *NGSS* and *CCSS* allow for students to learn in diverse ways, and the STEM Road Map curriculum modules emphasize that students can use a variety of tactics to complete the learning process. For example, students can use STEM Research Notebooks to record what they have learned during the various research activities. Notebook entries might include putting objectives in students' own words, compiling their prior learning on the topic, documenting new learning, providing proof of what they learned, and reflecting on what they felt successful doing and what they felt they still needed to work on. Perhaps students didn't realize that they were supposed to connect what they already knew with what they learned. They could record this and would be prepared in the next learning task to begin connecting prior learning with new learning.

SAFETY IN STEM

Student safety is a primary consideration in all subjects but is an area of particular concern in science, where students may interact with unfamiliar tools and materials that may pose additional safety risks. It is important to implement safety practices within the context of STEM investigations, whether in a classroom laboratory or in the field. When you keep safety in mind as a teacher, you avoid many potential issues with the lesson while also protecting your students.

STEM safety practices encompass things considered in the typical science classroom. Ensure that students are familiar with basic safety considerations, such as wearing protective equipment (e.g., safety glasses or goggles and latex-free gloves) and taking care with sharp objects, and know emergency exit procedures. Teachers should learn beforehand the locations of the safety eyewash, fume hood, fire extinguishers, and emergency shut-off switch in the classroom and how to use them. Also be aware of any school or district safety policies that are in place and apply those that align with the work being conducted in the lesson. It is important to review all safety procedures annually.

STEM investigations should always be supervised. Each lesson in the modules includes teacher guidelines for applicable safety procedures that should be followed. Before each investigation, teachers should go over these safety procedures with the student teams. Some STEM focus areas such as engineering require that students can demonstrate how to properly use equipment in the maker space before the teacher allows them to proceed with the lesson.

Information about classroom science safety, including a safety checklist for science classrooms, general lab safety recommendations, and links to other science safety resources, is available at the Council of State Science Supervisors (CSSS) website at *www.csss-science.org/safety.shtml*. The National Science Teachers Association (NSTA) provides a list of science rules and regulations, including standard operating procedures for lab safety, and a safety acknowledgement form for students and parents or guardians to sign. You can access this forum at *http://static.nsta.org/pdfs/SafetyInTheScienceClassroomLabAndField.pdf*. In addition, NSTA's Safety in the Science Classroom web page (*www.nsta.org/safety*) has numerous links to safety resources, including papers written by the NSTA Safety Advisory Board.

Disclaimer: The safety precautions for each activity are based on use of the recommended materials and instructions, legal safety standards, and better professional practices. Using alternative materials or procedures for these activities may jeopardize the level of safety and therefore is at the user's own risk. Further information regarding safety procedures can be found in other NSTA publications, such as the guide "Safety in the Science Classroom, Laboratory, or Field" (*http://static.nsta.org/pdfs/SafetyInTheScience Classroom.pdf*).

REFERENCES

Bandura, A. 1986. *Social foundations of thought and action: A social cognitive theory.* Englewood Cliffs, NJ: Prentice-Hall.

Barell, J. 2006. *Problem-based learning: An inquiry approach.* Thousand Oaks, CA: Corwin Press.

Benner, P. 1984. *From novice to expert: Excellence and power in clinical nursing practice.* Menlo Park, CA: Addison-Wesley Publishing Company.

Black, P., C. Harrison, C. Lee, B. Marshall, and D. Wiliam. 2003. *Assessment for learning: Putting it into practice.* Berkshire, UK: Open University Press.

Black, P., and D. Wiliam. 1998. Inside the black box: Raising standards through classroom assessment. *Phi Delta Kappan* 80 (2): 139–148.

Blumenfeld, P., E. Soloway, R. Marx, J. Krajcik, M. Guzdial, and A. Palincsar. 1991. Motivating project-based learning: Sustaining the doing, supporting learning. *Educational Psychologist* 26 (3): 369–398.

Brookhart, S. M., and A.J. Nitko. 2008. *Assessment and grading in classrooms.* Upper Saddle River, NJ: Pearson.

Bybee, R., J. Taylor, A. Gardner, P. Van Scotter, J. Carlson, A. Westbrook, and N. Landes. 2006. *The BSCS 5E instructional model: Origins and effectiveness. http://science.education.nih.gov/houseofreps. nsf/b82d55fa138783c2852572c9004f5566/$FILE/Appendix?D.pdf.*

Eliason, C. F., and L. T. Jenkins. 2012. *A practical guide to early childhood curriculum.* 9th ed. New York: Merrill.

Johnson, C. 2003. Bioterrorism is real-world science: Inquiry-based simulation mirrors real life. *Science Scope* 27 (3): 19–23.

Krajcik, J., and P. Blumenfeld. 2006. Project-based learning. In *The Cambridge handbook of the learning sciences,* ed. R. Keith Sawyer, 317–334. New York: Cambridge University Press.

Lambros, A. 2004. *Problem-based learning in middle and high school classrooms: A teacher's guide to implementation.* Thousand Oaks, CA: Corwin Press.

Milner, A. R., and T. Sondergeld. 2015. Gifted urban middle school students: The inquiry continuum and the nature of science. *National Journal of Urban Education and Practice* 8 (3): 442–461.

Morrison, V., and A. R. Milner. 2014. Literacy in support of science: A closer look at cross-curricular instructional practice. *Michigan Reading Journal* 46 (2): 42–56.

National Association for the Education of Young Children (NAEYC). 2016. Developmentally appropriate practice position statements. *www.naeyc.org/positionstatements/dap.*

Peters, E. E. 2010. Shifting to a student-centered science classroom: An exploration of teacher and student changes in perceptions and practices. *Journal of Science Teacher Education* 21 (3): 329–349.

Peters, E. E. 2012. Developing content knowledge in students through explicit teaching of the nature of science: Influences of goal setting and self-monitoring. *Science and Education* 21 (6): 881–898.

Peters, E. E., and A. Kitsantas. 2010. The effect of nature of science metacognitive prompts on science students' content and nature of science knowledge, metacognition, and self-regulatory efficacy. *School Science and Mathematics* 110: 382–396.

Popham, W. J. 2013. *Classroom assessment: What teachers need to know.* 7th ed. Upper Saddle River, NJ: Pearson.

Ritchhart, R., M. Church, and K. Morrison. 2011. *Making thinking visible: How to promote engagement, understanding, and independence for all learners.* San Francisco, CA: Jossey-Bass.

Sondergeld, T. A., C. A. Bell, and D. M. Leusner. 2010. Understanding how teachers engage in formative assessment. *Teaching and Learning* 24 (2): 72–86.

Zimmerman, B. J. 2000. Attaining self-regulation: A social-cognitive perspective. In *Handbook of self-regulation,* ed. M. Boekaerts, P. Pintrich, and M. Zeidner, 13–39. San Diego: Academic Press.

PART 2

TRANSPORTATION IN THE FUTURE

STEM ROAD MAP MODULE

TRANSPORTATION IN THE FUTURE MODULE OVERVIEW

Janet B. Walton, Sandy Watkins, Carla C. Johnson, and Erin E. Peters-Burton

THEME: Innovation and Progress

LEAD DISCIPLINES: Social Studies and Science

MODULE SUMMARY

In this module, students learn about the geography of the continental United States, explore the role of trains in the nation's development, and consider train travel in the context of the 21st century. Students develop conceptual understanding of innovations in train technology, with a focus on magnetic levitation (maglev) trains. Inquiry activities in science allow students to gain an understanding of magnetic interactions. Using mathematics, students learn to calculate distances and time intervals. Students apply this knowledge in the Maglevacation Train Challenge, working collaboratively using the engineering design process (EDP) to create a prototype train that can safely carry passengers a given distance and then to prepare a presentation that includes information about their destination and the design features of their train (adapted from Capobianco et al. 2015).

ESTABLISHED GOALS AND OBJECTIVES

At the conclusion of this module, students will be able to do the following:

- Apply their understanding of U.S. geography to locate a destination and calculate distances

- Use map skills to identify geographic features including rivers, mountains, and oceans

- Understand the relationship between the development of the rail system and the progressive development of the United States

- Apply their conceptual understanding of the history of train technology and its role in train efficiency, performance, and safety to solving a design challenge

- Apply their understanding of magnetism to create a maglev vehicle

- Understand that magnets can play an important role in transportation innovations

- Apply mathematics skills to calculate distances, speeds, and time intervals

- Demonstrate their understanding of the EDP by using it to collaboratively complete assigned tasks

- Discuss geographic and demographic features of a location in the continental United States

- Present their projects in a clear, concise presentation format

CHALLENGE OR PROBLEM FOR STUDENTS TO SOLVE: THE MAGLEVACATION TRAIN CHALLENGE

Student teams are challenged to each choose a vacation destination, research that destination, and create a prototype Maglevacation Train to carry passengers to that destination. Then, each team creates a video presentation to provide information about its destination and train to a fictional client, including the following:

- Geographic, climatic, and cultural information about the destination

- A review of the design features of the prototype train

- A demonstration of the prototype's performance

- A persuasive argument about why travelers should choose the team's destination and train

Driving Question: How can we create a plan and build a prototype for a maglev train to carry passengers to a vacation destination?

CONTENT STANDARDS ADDRESSED IN THIS STEM ROAD MAP MODULE

A full listing with descriptions of the standards this module addresses can be found in the appendix. Listings of the particular standards addressed within lessons are provided in a table for each lesson in Chapter 4.

STEM RESEARCH NOTEBOOK

Each student should maintain a STEM Research Notebook, which will serve as a place for students to organize their work throughout this module (see p. 26 for more general discussion on setup and use of this notebook). All written work in the module should be included in the notebook, including records of students' thoughts and ideas, fictional accounts based on the concepts in the module, and records of student progress through the EDP. The notebooks may be maintained across subject areas, giving students the opportunity to see that although their classes may be separated during the school day, the knowledge they gain is connected.

Each lesson in this module includes student handouts that should be kept in the STEM Research Notebooks after completion, as well as a prompt to which students should respond in their notebooks. Students will have the opportunity to create covers and tables of contents for their Research Notebooks in Lesson 1. You may also wish to have students include the STEM Research Notebook Guidelines student handout on p. 26 in their notebooks.

Emphasize to students the importance of organizing all information in a Research Notebook. Explain to them that scientists and other researchers maintain detailed Research Notebooks in their work. These notebooks, which are crucial to researchers' work because they contain critical information and track the researchers' progress, are often considered legal documents for scientists who are pursuing patents or wish to provide proof of their discovery process.

STUDENT HANDOUT

STEM RESEARCH NOTEBOOK GUIDELINES

STEM professionals record their ideas, inventions, experiments, questions, observations, and other work details in notebooks so that they can use these notebooks to help them think about their projects and the problems they are trying to solve. You will each keep a STEM Research Notebook during this module that is like the notebooks that STEM professionals use. In this notebook, you will include all your work and notes about ideas you have. The notebook will help you connect your daily work with the big problem or challenge you are working to solve.

It is important that you organize your notebook entries under the following headings:

1. **Chapter Topic or Title of Problem or Challenge:** You will start a new chapter in your STEM Research Notebook for each new module. This heading is the topic or title of the big problem or challenge that your team is working to solve in this module.

2. **Date and Topic of Lesson Activity for the Day:** Each day, you will begin your daily entry by writing the date and the day's lesson topic at the top of a new page. Write the page number both on the page and in the table of contents.

3. **Information Gathered From Research:** This is information you find from outside resources such as websites or books.

4. **Information Gained From Class or Discussions With Team Members:** This information includes any notes you take in class and notes about things your team discusses. You can include drawings of your ideas here, too.

5. **New Data Collected From Investigations:** This includes data gathered from experiments, investigations, and activities in class.

6. **Documents:** These are handouts and other resources you may receive in class that will help you solve your big problem or challenge. Paste or staple these documents in your STEM Research Notebook for safekeeping and easy access later.

7. **Personal Reflections:** Here, you record your own thoughts and ideas on what you are learning.

8. **Lesson Prompts:** These are questions or statements that your teacher assigns you within each lesson to help you solve your big problem or challenge. You will respond to the prompts in your notebook.

9. **Other Items:** This section includes any other items your teacher gives you or other ideas or questions you may have.

MODULE LAUNCH

To launch the module, have students engage in a group discussion about train travel to activate prior knowledge, share personal experiences, and highlight perceptions about this mode of travel. After the discussion, the class views a video clip about transportation that emphasizes high-speed trains and the advanced technology associated with this mode of travel. (Relevant videos can be found on YouTube by searching for "future of transportation technologies"; one example is "Hyperloop and Future Transport Technology" at *www.youtube.com/watch?v=YHiKjJEFY6A*.) After viewing the video, the students revisit their earlier discussion and reflect on what surprised them about train travel in this video and what they learned.

Tell students that as part of their challenge in this module, they will be acting as design engineers to create a model or prototype of a train that can carry passengers from their hometown to a vacation spot of their choice as quickly and safely as possible and will also be acting as travel agents to provide information about a vacation spot.

PREREQUISITE SKILLS FOR THE MODULE

Students enter this module with a wide range of preexisting skills, information, and knowledge. Table 3.1 (p. 28) provides an overview of prerequisite skills and knowledge that students are expected to apply in this module, along with examples of how they apply this knowledge throughout the module. Differentiation strategies are also provided for students who may need additional support in acquiring or applying this knowledge.

Table 3.1. Prerequisite Key Knowledge and Examples of Applications and Differentiation Strategies

Prerequisite Key Knowledge	Application of Knowledge by Students	Differentiation for Students Needing Additional Knowledge
Measurement skills: • Distance • Time	Measurement skills: • Measure distances using standard units. • Use timetables and clocks to measure arrival and departure times for trains to the nearest minute.	Measurement skills: • Provide students with opportunities to practice measuring distances using various units and measuring time to the nearest minute. • Provide students with additional content, including textbook support, teacher instruction, and online videos for telling time to the nearest minute.
Inquiry skills: • Ask questions, make logical predictions, plan investigations, assess and address safety issues, and represent data. • Use senses and simple tools to make observations. • Communicate interest in simple phenomena and plan for simple investigations. • Communicate understanding of simple data using age-appropriate vocabulary.	Inquiry skills: • Select and use appropriate tools, simple equipment, and safety equipment to conduct an investigation. • Identify tools needed to investigate specific questions. • Maintain a STEM Research Notebook that includes observations, data, diagrams, and explanations. • Analyze and communicate findings from multiple investigations of similar phenomena to reach a conclusion.	Inquiry skills: • Select, model, and use appropriate tools and simple equipment to help students conduct an investigation. • Provide samples of a STEM Research Notebook. • Scaffold student efforts to organize data into tables, graphs, drawings, or diagrams by providing step-by-step instructions. • Use classroom discussions to identify specific investigations that could be used to answer a particular question and identify reasons for this choice.
Numbers and operations: • Add and subtract numbers within 1,000. • Multiply and divide whole numbers.	Numbers and operations: • Engage in activities that involve finding sums of numbers within 1,000. • Calculate distances using a map scale. • Calculate speeds in units of miles per hour.	Numbers and operations: • Review and provide models of adding and subtracting within 1,000 using the standard algorithm. • Review multiplication and division and provide examples of map scale and speed calculations.

Prerequisite Key Knowledge	Application of Knowledge by Students	Differentiation for Students Needing Additional Knowledge
Reading: • Use information gained from the illustrations and words in a print or digital text to demonstrate understanding of the connection between a series of historical events, scientific ideas or concepts, or steps in technical procedures in a text.	Reading: • Describe the relationship between a series of historical events and train travel using language that pertains to time, sequence, and cause and effect.	Reading: • Provide reading strategies to support comprehension of nonfiction texts, including activating prior knowledge, previewing text by skimming content and scanning images, and rereading.
Writing: • Write informative/explanatory and narrative texts in which students introduce a topic, use facts and definitions to develop points, and provide a concluding statement or section.	Writing: • Write informative/explanatory texts to examine a topic and convey ideas and information clearly. • Write narratives to develop real or imagined experiences or events using effective technique, descriptive details, and clear event sequences.	Writing: • Provide a template for writing informative/explanatory texts to scaffold student writing exercises. • Provide writing organizer handouts to scaffold student work in describing details and clarifying event sequence.
Speaking and listening: • Participate in collaborative conversations using appropriate language and listening skills. • Tell a story or recount an experience with appropriate facts and relevant, descriptive details, speaking audibly in coherent sentences.	Speaking and listening: • Engage in a number of collaborative discussions and presentations in which students need to provide evidence and speak persuasively. • Present factual information to an audience.	Speaking and listening: • Scaffold student understanding of speaking skills by providing examples of appropriate language and presentations, with an emphasis on presentation techniques and language use. • Provide handouts to support organization of appropriate facts and relevant descriptive details for presentations.

POTENTIAL STEM MISCONCEPTIONS

Students enter the classroom with a wide variety of prior knowledge and ideas, so it is important to be alert to misconceptions, or inappropriate understandings of foundational knowledge. These misconceptions can be classified as one of several types: "preconceived notions," opinions based on popular beliefs or understandings; "nonscientific beliefs," knowledge students have gained about science from sources outside the

scientific community; "conceptual misunderstandings," incorrect conceptual models based on incomplete understanding of concepts; "vernacular misconceptions," misunderstandings of words based on their common use versus their scientific use; and "factual misconceptions," incorrect or imprecise knowledge learned in early life that remains unchallenged (NRC 1997, p. 28). Misconceptions must be addressed and dismantled in order for students to reconstruct their knowledge, and therefore teachers should be prepared to take the following steps:

- *Identify students' misconceptions.*

- *Provide a forum for students to confront their misconceptions.*

- *Help students reconstruct and internalize their knowledge, based on scientific models. (NRC 1997, p. 29)*

Keeley and Harrington (2010) recommend using diagnostic tools such as probes and formative assessment to identify and confront student misconceptions and begin the process of reconstructing student knowledge. Keeley and Harrington's *Uncovering Student Ideas in Science* series contains probes targeted toward uncovering student misconceptions in a variety of areas. In particular, Volumes 1 and 2 of *Uncovering Student Ideas in Physical Science* (Keeley and Harrington 2010, 2014), about force/motion and electricity/magnetism, may be useful resources for addressing student misconceptions in this module.

Some commonly held misconceptions specific to lesson content are provided with each lesson so that you can be alert for student misunderstanding of the science concepts presented and used during this module. The American Association for the Advancement of Science has also identified misconceptions that students frequently hold regarding various science concepts (see the links at *http://assessment.aaas.org/topics*).

SRL PROCESS COMPONENTS

Table 3.2 illustrates some of the activities in the Transportation in the Future module and how they align to the SRL process before, during, and after learning.

STRATEGIES FOR DIFFERENTIATING INSTRUCTION WITHIN THIS MODULE

For the purposes of this curriculum module, differentiated instruction is conceptualized as a way to tailor instruction—including process, content, and product—to various student needs in your class. A number of differentiation strategies are integrated into lessons across the module. The problem- and project-based learning approach used in the lessons is designed to address students' multiple intelligences by providing a variety of entry points and methods to investigate the key concepts in the module (for example,

Table 3.2. SRL Process Components

Learning Process Components	Examples From Transportation in the Future Module	Lesson Number and Learning Component
BEFORE LEARNING		
Motivates students	Students choose a destination for the maglev train they will create.	Lesson 1, Introductory Activity/Engagement
Evokes prior learning	Students engage in group discussions about their personal experiences with trains and other modes of travel and reflect on the availability of trains for passenger travel in their geographic area.	Lesson 1, Introductory Activity/Engagement
DURING LEARNING		
Focuses on important features	Students gain experience in using magnets as a propulsion system in the Magnificent Magnet Match mini design challenge.	Lesson 2, Activity/ Exploration
Helps students monitor their progress	Students create a STEM Research Notebook entry reflecting on how their team used the engineering design process, with a focus on things that went well, challenges, and teamwork.	Lesson 2, Elaboration/ Application of Knowledge
AFTER LEARNING		
Evaluates learning	Students receive feedback on rubrics for their Maglevacation Train design and their presentations.	Lesson 4, Assessment
Takes account of what worked and what did not work	Students examine their prior work in their STEM Research Notebooks to determine what they learned over the course of the lesson and how this learning can be applied to creating a solution for the module challenge.	Lesson 4, Activity/ Exploration

investigating magnetism via scientific inquiry, literature, journaling, and collaborative design). Differentiation strategies for students needing support in prerequisite knowledge can be found in Table 3.1. You are encouraged to use information gained about student prior knowledge during introductory activities and discussions to inform your instructional differentiation. Strategies incorporated into this lesson include flexible grouping, varied environmental learning contexts, assessments, compacting, and tiered assignments and scaffolding.

Flexible Grouping: Students work collaboratively in a variety of activities throughout this module. Grouping strategies you might employ include student-led grouping, grouping students according to ability level, grouping students randomly, or grouping them so that students in each group have complementary strengths (for instance, one student might be strong in mathematics, another in art, and another in writing). You may also choose to group students based on their prior knowledge about trains and magnetism. For Lesson 2, you may choose to maintain the same student groupings as in Lesson 1 or regroup students according to another of the strategies described here. Students begin to use the EDP during the Magnificent Magnet Match in Lesson 2 and continue to use the EDP throughout the module and for their final challenge. You may therefore wish to consider grouping students in Lesson 2 into design teams on which they will remain throughout the module.

Varied Environmental Learning Contexts: Students have the opportunity to learn in various contexts throughout the module, including alone, in groups, in quiet reading and research-oriented activities, and in active learning through inquiry and design activities. In addition, students learn in a variety of ways, including through doing inquiry activities, journaling, reading fiction and nonfiction texts, watching videos, participating in class discussion, and conducting web-based research.

Assessments: Students are assessed in a variety of ways throughout the module, including individual and collaborative formative and summative assessments. Students have the opportunity to produce work via written text, oral and media presentations, and modeling. You may choose to provide students with additional choices of media for their products (for example, PowerPoint presentations, posters, or student-created websites or blogs).

Compacting: Based on student prior knowledge, you may wish to adjust instructional activities for students who exhibit prior mastery of a learning objective. For instance, if some students exhibit mastery of calculating map distances and speeds in mathematics in Lesson 1, you may wish to limit the amount of time they spend practicing these skills and instead introduce various units of measurement and unit conversions to these students or introduce the concept of map scale and proportions with associated activities.

Tiered Assignments and Scaffolding: Based on your awareness of student ability, understanding of concepts, and mastery of skills, you may wish to provide students with variations on activities by adding complexity to assignments or providing more or fewer learning supports for activities throughout the module. For instance, some students may need additional support in identifying key search words and phrases for web-based research or may benefit from cloze sentence handouts to enhance vocabulary understanding. Other students may benefit from expanded reading selections and additional reflective writing or from working with manipulatives and other visual representations

of mathematical concepts. You may also work with your school librarian to compile a set of topical resources at a variety of reading levels.

STRATEGIES FOR ENGLISH LANGUAGE LEARNERS

Students who are developing proficiency in English language skills require additional supports to simultaneously learn academic content and the specialized language associated with specific content areas. The World-Class Instructional Design and Assessment Consortium (WIDA) has created a framework for providing support to these students and makes available rubrics and guidance on differentiating instructional materials for English language learners (ELLs) (see *www.wida.us/get.aspx?id=7*). In particular, ELL students may benefit from additional sensory supports such as images, physical modeling, and graphic representations of module content, as well as interactive support through collaborative work. This module incorporates a variety of sensory supports and offers ongoing opportunities for ELL students to work with collaboratively. The focus in this module on various modes of transportation and high-speed trains in a global context affords opportunities to access the culturally diverse experiences of ELL students in the classroom.

Teachers differentiating instruction for ELL students should carefully consider the needs of these students as they introduce and use academic language in various language domains (listening, speaking, reading, and writing) throughout this module. To adequately differentiate instruction for ELL students, teachers should have an understanding of the proficiency level of each student. The following five overarching preK–5 WIDA learning standards are relevant to this module:

- Standard 1: Social and Instructional language. Focus on social behavior in group work and class discussions.

- Standard 2: The language of Language Arts. Focus on forms of print, elements of text, picture books, comprehension strategies, main ideas and details, persuasive language, creation of informational text, and editing and revision.

- Standard 3: The language of Mathematics. Focus on numbers and operations, patterns, number sense, measurement, and strategies for problem solving.

- Standard 4: The language of Science. Focus on safety practices, magnetism, energy sources, scientific process, and scientific inquiry.

- Standard 5: The language of Social Studies. Focus on change from past to present, historical events, resources, transportation, map reading, and location of objects and places.

Table 3.3. Desired Outcomes and Evidence of Success in Achieving Identified Outcomes

Desired Outcomes	Evidence of Success	
	Performance tasks	Other measures
Students can apply an understanding of geography, mapping, and various science and mathematics concepts to complete small group projects and individual tasks related to the projects within the unit.	• Students maintain STEM Research Notebooks that contain designs, research notes, evidence of collaboration, and ELA-related work. • Students design a working prototype. • Student teams research and present information on a destination of their choice. • Students are able to discuss how they applied their understanding of concepts introduced in the module to their designs (individual and team) and presentations. • Students are assessed using project rubrics that focus on content and application of skills related to the academic content.	• Student collaboration is assessed using a collaboration rubric.

SAFETY CONSIDERATIONS FOR THE ACTIVITIES IN THIS MODULE

This module's science component focuses on magnets. Ensure that no students have health conditions that could be affected by use of magnets (e.g., pacemakers). All laboratory occupants must wear safety glasses or goggles during all phases of inquiry activities (setup, hands-on investigation, and takedown). Students should be instructed not to blow on or throw iron filings. For more general safety guidelines, see the section on Safety in STEM in Chapter 2 (p. 18).

DESIRED OUTCOMES AND MONITORING SUCCESS

The desired outcomes for this module are outlined in Table 3.3, along with suggested ways to gather evidence to monitor student success. For more specific details on desired outcomes, see the Established Goals and Objectives sections for the module and the individual lessons.

ASSESSMENT PLAN OVERVIEW AND MAP

Table 3.4 provides an overview of the major group and individual *products* and *deliverables,* or things that constitute the assessment for this module. See Table 3.5 (p. 36) for a full assessment map of formative and summative assessments in this module.

Table 3.4. Major Products and Deliverables in Lead Disciplines for Groups and Individuals

Lesson	Major Group Products and Deliverables	Major Individual Products and Deliverables
1	• Map Me! group decision and geographic information about vacation destination	• Map Me! handouts and maps • Magnet preassessment • Magnetic or Not? data sheet handout • STEM Research Notebook prompt
2	• My Magnetic Question presentation • Magnificent Magnet Match boat design	• My Magnetic Question research question • Magnificent Magnet Match Engineer It! handouts • STEM Research Notebook prompt
3	• Expert group contribution to Planes, Trains, and Automobiles class chart • Let's Go! presentation and budget	• Electromagnetic Demonstration handouts and data • Planes, Trains, and Automobiles expert group graphic organizer • Let's Go! handouts • Balloon Car handouts and data • STEM Research Notebook prompt
4	• Maglevacation Train Challenge presentation and train design	• Maglevacation Train Challenge student packet pages • STEM Research Notebook prompt

Table 3.5. Assessment Map for Transportation in the Future Module

Lesson	Assessment	Group/ Individual	Formative/ Summative	Lesson Objective Assessed
1	Map Me! *handouts*	Individual	Formative	• Use a map to identify student's region, state, county, and town. • Identify the geographic features of student's hometown and state.
1	Magnetic or Not? *handouts*	Individual	Formative	• Describe and demonstrate the ways magnetic poles attract and repel one another.
1	Magnet Magic *handouts*	Individual	Formative	• Describe and demonstrate the ways magnetic poles attract and repel one another.
1	My Magnetic Question *STEM Research Notebook entries*	Individual	Formative	• Formulate a testable research question. • Design an investigation and draw conclusions from that investigation
1	STEM Research Notebook *prompt*	Individual	Formative	• Calculate distances on a map using the map scale.
2	Riding the Rails *handouts*	Individual	Formative	• Demonstrate ability to locate various U.S. locations on a map and identify geographic features of those areas.
2	Magnificent Magnet Match *boat design*	Group	Formative	• Apply understanding of magnetism and the EDP to a group design challenge.
2	Magnificent Magnet Match *EDP handouts*	Individual	Formative	• Apply understanding of magnetism to a group design challenge. • Demonstrate understanding of the EDP.
2	Evidence of collaboration *collaboration rubric*	Individual	Formative	• Use the EDP to work collaboratively.
2	STEM Research Notebook *prompt*	Individual	Formative	• Identify and discuss various features and uses of maps.
3	Planes, Trains, and Automobiles *graphic organizer*	Individual	Formative	• Use understanding of the advantages and disadvantages of various modes of transportation to make recommendations about the preferred method of travel to a particular destination.

Lesson	Assessment	Group/Individual	Formative/Summative	Lesson Objective Assessed
3	Let's Go! *presentation and budget*	Group	Summative	• Use understanding of geography and maps to predict climate conditions in various locations around the United States. • Apply understanding of budgets to create a travel budget. • Apply findings from research to calculate costs per mile. • Create a compelling presentation using technology to highlight students' findings. • Create a persuasive argument for visiting the chosen vacation destination.
3	Let's Go! *handouts and graphic organizer*	Individual	Summative	• Use understanding of geography and maps to predict climate conditions in various locations around the United States. • Apply understanding of budgets to create a travel budget. • Create a persuasive argument for visiting the chosen vacation destination. • Apply findings from research to calculate costs per mile.
3	Electromagnetic Demonstration *handouts*	Individual	Formative	• Demonstrate a conceptual understanding of electromagnets. • Identify independent and dependent variables. • Formulate hypotheses and test these hypotheses.
3	Balloon Car *handouts*	Individual	Formative	• Understand that objects must have a force applied to them to initiate movement. • Understand and demonstrate that the weight of an object affects how much force is needed to initiate motion. • Formulate hypotheses and test these hypotheses.

(continued)

Table 3.5. (*continued*)

Lesson	Assessment	Group/ Individual	Formative/ Summative	Lesson Objective Assessed
3	STEM Research Notebook *prompt*	Individual	Formative	• Understand that objects must have a force applied to them to initiate movement and demonstrate this understanding by creating a vehicle propelled by air leaving a balloon. • Discuss how magnets are used in maglev trains and predict how this could be used to create student's own prototype maglev vehicle.
4	Maglevacation Train Challenge *student packet*	Individual	Summative	• Apply understanding of mapping and geography to create a persuasive presentation about a travel destination. • Apply understanding of how speed and weight affect acceleration to create the fastest speeds possible over a short distance.
4	Prototype design *rubric*	Group	Summative	• Apply understanding of speed and how weight affects acceleration to create the fastest speeds possible over a short distance. • Apply understanding of magnetism to create a prototype maglev train.
4	Video presentation *rubric*	Group	Summative	• Synthesize learning throughout the module to create a presentation appropriate for the audience.
4	Collaboration *rubric*	Individual	Summative	• Apply understanding of the EDP to work collaboratively to create a solution to a challenge.

MODULE TIMELINE

Tables 3.6–3.10 (pp. 39–40) provide lesson timelines for each week of the module. These timelines are provided for general guidance only and are based on class times of approximately 45 minutes.

Table 3.6. STEM Road Map Module Schedule for Week One

Day 1	Day 2	Day 3	Day 4	Day 5
Lesson 1: Maglevs, Maps, and Magnets • Launch the module. Activate student prior knowledge of trains through discussions and video clips. • Conduct preassessment on magnets and magnetism.	*Lesson 1: Maglevs, Maps, and Magnets* • Students form design teams and formulate a list of potential destinations, practice map reading skills, and begin Map Me! activity. • Introduce physical properties of matter.	*Lesson 1: Maglevs, Maps, and Magnets* • Continue mapping activities. • Introduce magnetism Magnetic or Not? activity.	*Lesson 1: Maglevs, Maps, and Magnets* • Continue mapping and map reading activities. • Introduce Magnet Magic activity.	*Lesson 2: Trains Through Time* • Introduce Transcontinental Railroad. • Introduce Riding the Rails activity. • Introduce testable questions (My Magnetic Question).

Table 3.7. STEM Road Map Module Schedule for Week Two

Day 6	Day 7	Day 8	Day 9	Day 10
Lesson 2: Trains Through Time • Continue Riding the Rails activity. • Do My Magnetic Question activity. • Introduce biomimicry.	*Lesson 2: Trains Through Time* • Introduce EDP. • Begin Magnificent Magnet Match activity. • Introduce timetables and elapsed time.	*Lesson 2: Trains Through Time* • Continue Magnificent Magnet Match. • Investigate availability of passenger rail travel and time zones.	*Lesson 3: Portable People* • Explore various forms of transportation and locations of maglev trains. • Explore budgeting. • Conduct Electromagnetic Demonstration.	*Lesson 3: Portable People* • Students form expert groups to investigate a travel topic across three modes of transportation in the Planes, Trains, and Automobiles activity.

Table 3.8. STEM Road Map Module Schedule for Week Three

Day 11	Day 12	Day 13	Day 14	Day 15
Lesson 3: Portable People • Continue Planes, Trains, and Automobiles activity. • Introduce Balloon Car activity.	*Lesson 3: Portable People* • Design Teams begin Let's Go! activity (background research). • Complete Balloon Car activity.	*Lesson 3: Portable People* • Continue Let's Go! activity (construct budget).	*Lesson 3: Portable People* • Continue Let's Go! activity.	*Lesson 3: Portable People* • Finish Let's Go! activity (share student presentations).

Table 3.9. STEM Road Map Module Schedule for Week Four

Day 16	Day 17	Day 18	Day 19	Day 20
Lesson 4: Speeding Ahead— The Maglevacation Train Challenge • Introduce challenge and challenge parameters. • Begin prototype design.	*Lesson 4: Speeding Ahead— The Maglevacation Train Challenge* • Complete prototype maglev train design and build.	*Lesson 4: Speeding Ahead— The Maglevacation Train Challenge* • Test and redesign prototype. • Record speeds.	*Lesson 4: Speeding Ahead— The Maglevacation Train Challenge* • Compile and organize information for presentation.	*Lesson 4: Speeding Ahead— The Maglevacation Train Challenge* • Compile and organize information for presentation.

Table 3.10. STEM Road Map Module Schedule for Week Five

Day 21	Day 22	Day 23	Day 24	Day 25
Lesson 4: Speeding Ahead— The Maglevacation Train Challenge • Complete organization of presentation and practice.	*Lesson 4: Speeding Ahead— The Maglevacation Train Challenge* • Create videos.	*Lesson 4: Speeding Ahead— The Maglevacation Train Challenge* • Share videos with audience.	• These days are left open to accommodate any lessons that may have taken longer than anticipated. If the module is completed by Day23, options include taking a field trip to a local train station or museum, having a railway worker visit the classroom, or showing student videos to other classes in the school.	

3

RESOURCES

Teachers have the option to co-teach portions of this unit and may want to combine classes for activities such as mathematical modeling, geometric investigations, discussing social influences, or conducting research. The media specialist can help teachers locate resources for students to view and read about the history of transportation and provide technical help with spreadsheets, timeline software, and multimedia production software. Special educators and reading specialists can help find supplemental sources for students needing extra support in reading and writing. Additional resources may be found online. Community resources for this module may include travel agents, engineers, school administrators, and parents.

REFERENCES

Capobianco, B. M., C. Parker, A. Laurier, and J. Rankin. 2015. The STEM road map for grades 3–5. In *STEM Road Map: A framework for integrated STEM education,* ed. C. C. Johnson, E. E. Peters-Burton, and T. J. Moore, 68–95. New York: Routledge. *www.routledge.com/products/9781138804234.*

Keeley, P., and R. Harrington. 2010. *Uncovering student ideas in physical science. Vol. 1, 45 new force and motion assessment probes.* Arlington, VA: NSTA Press.

Keeley, P., and R. Harrington. 2014. *Uncovering student ideas in physical science. Vol. 2, 39 new electricity and magnetism formative assessment probes.* Arlington, VA: NSTA Press.

National Research Council (NRC). 1997. *Science teaching reconsidered: A handbook.* Washington, DC: National Academies Press.

World-Class Instructional Design and Assessment Consortium (WIDA). 2012. 2012 Amplification of the English language development standards: Kindergarten–grade 12. *www.wida.us/standards/eld.aspx.*

TRANSPORTATION IN THE FUTURE LESSON PLANS

Janet B. Walton, Sandy Watkins, Carla C. Johnson, and Erin E. Peters-Burton

Lesson Plan 1: Maglevs, Maps, and Magnets

This lesson introduces students to the module and the culminating challenge for the module, the Maglevacation Train Challenge. Discussions and video clips are designed to ignite students' curiosity and activate prior knowledge about trains. Social studies activities focus on map reading and a basic understanding of U.S. geography that students apply in choosing a destination for their Maglevacation Trains. In science, students investigate magnets and reflect on how what they learn about magnetism that may be useful as they approach their challenge.

ESSENTIAL QUESTIONS

- How can we use maps to determine a location?

- What region of the United States do we live in?

- How have technological advances affected the performance of trains?

- How do magnets interact with one another and with other materials?

- Why do magnets attract and repel one another?

- What is a testable question?

ESTABLISHED GOALS AND OBJECTIVES

At the conclusion of this lesson, students will be able to do the following:

- Use a map to identify their region, state, county, and town

- Identify and discuss various features of and uses of maps

- Identify the geographic features of their hometown and state

- Calculate distances on a map using the map scale

- Understand that speed affects the time it takes to travel a given distance
- Describe how magnets interact with various materials
- Describe and demonstrate the ways magnetic poles attract and repel one another
- Formulate a testable research question
- Design an investigation and draw conclusions from that investigation

TIME REQUIRED

- 4 days (approximately 45 minutes each day; see Table 3.6)

MATERIALS

Required Materials for Lesson 1

- STEM Research Notebooks (1 per student; see p. 26 for STEM Research Notebook student handout)
- Access to internet technology for showing video clips and for student research
- Classroom map of the United States
- Chart paper (for Know/Want to Know/Learned [KWL] charts)
- Ball
- Magnet
- Assorted materials with varying magnetic properties (e.g., nails, paper clips, pieces of wood, metal washers, fabric, aluminum foil)
- Handouts (attached at the end of this lesson)

Additional Materials for Map Me!

- U.S. road map (1 per group)
- Ruler
- String (3 feet per group)
- Map Me! handouts (1 set per student; attached at the end of this lesson)
- Colored pencils (1 set per student)
- List of potential destinations and student-created lists of information about the destinations (created in Introductory Activity/Engagement phase)

Additional Materials for Magnetic or Not? (per group of 2–3 students unless otherwise indicated)

- Safety glasses or goggles (1 pair per student)
- Pencil
- Piece of string (about 18 inches long)
- Magnet that can easily be tied onto the end of a string
- Ruler
- Magnetic or Not? handout (1 per student; attached at the end of this lesson)

Additional Materials for Magnet Magic (per pair of students unless otherwise indicated)

- Safety glasses or goggles (1 pair per student)
- Nonlatex gloves (1 pair per student)
- 2 bar magnets
- 2 small, round magnets (must be small enough to fit inside the lines of the maze)
- Resealable plastic sandwich bag*
- 3 × 5 index card*
- 1 tsp. iron filings*
- Duct tape to seal bags
- 12 oz. clear plastic cup
- Water to fill cup
- Magnet Magic Maze (attached at the end of this lesson)
- Magnet Magic handouts (1 set per student; attached at the end of this lesson)

*See Preparation for Lesson 1 on page 52 for more details.

CONTENT STANDARDS AND KEY VOCABULARY

Table 4.1 (p. 46) lists the content standards from the *Next Generation Science Standards* (*NGSS*), *Common Core State Standards* (*CCSS*), and Framework for 21st Century Learning that this lesson addresses, and Table 4.2 (p. 48) presents the key vocabulary. Vocabulary terms are provided for both teacher and student use. Teachers may choose to introduce some or all of the terms to students.

Table 4.1. Content Standards Addressed in STEM Road Map Module Lesson 1

NEXT GENERATION SCIENCE STANDARDS

PERFORMANCE EXPECTATION

3-PS2-3. Ask questions to determine cause-and-effect relationships of electric or magnetic interactions between two objects not in contact with each other.

SCIENCE AND ENGINEERING PRACTICE

Asking Questions and Defining Problems
- Ask questions about what would happen if a variable is changed.

DISCIPLINARY CORE IDEAS

PS2.A: Forces and Motion
- The patterns of an object's motion in various situations can be observed and measured; when that past motion exhibits a regular pattern, future motion can be predicted from it.

PS2.B: Types of Interactions
- Electric and magnetic forces between a pair of objects do not require that the objects be in contact. The sizes of the forces in each situation depend on the properties of the objects and their distances apart and, for forces between two magnets, on their orientation relative to each other.

CROSSCUTTING CONCEPTS

Cause and Effect
- Cause and effect relationships are routinely identified, tested, and used to explain change.

Influence of Science, Engineering, and Technology on Society and the Natural World
- People's needs and wants change over time, as do their demands for new and improved technologies.

COMMON CORE STATE STANDARDS FOR MATHEMATICS

MATHEMATICAL PRACTICES
- MP1. Make sense of problems and persevere in solving them.
- MP2. Reason abstractly and quantitatively.
- MP4. Model with mathematics.
- MP5. Use appropriate tools strategically.
- MP6. Attend to precision.

MATHEMATICAL CONTENT
- 3.NBT.A.2. Fluently add and subtract within 1000 using strategies and algorithms based on place value, properties of operations, and/or the relationship between addition and subtraction.
- 3.MD.B.4. Generate measurement data by measuring lengths using rulers marked with halves and fourths of an inch. Show the data by making a line plot, where the horizontal scale is marked off in appropriate units—whole numbers, halves, or quarters.

COMMON CORE STATE STANDARDS FOR ENGLISH LANGUAGE ARTS

READING STANDARDS

- RI.3.1. Ask and answer questions to demonstrate understanding of a text, referring explicitly to the text as the basis for the answers.

- RI.3.3. Describe the relationship between a series of historical events, scientific ideas or concepts, or steps in technical procedures in a text, using language that pertains to time, sequence, and cause/effect.

WRITING STANDARD

- W.3.2. Write informative/explanatory texts to examine a topic and convey ideas and information clearly.

SPEAKING AND LISTENING STANDARDS

- SL.3.1. Engage effectively in a range of collaborative discussions (one-on-one, in groups, and teacher-led) with diverse partners on *grade 3 topics and texts,* building on others' ideas and expressing their own clearly.

- SL3.1.D. Explain their ideas and understanding in light of the discussion.

- SL3.3. Ask and answer questions about information from a speaker, offering appropriate elaboration and detail.

FRAMEWORK FOR 21ST CENTURY LEARNING

Interdisciplinary themes (inventions, history, engineering design process, and progress); Learning and Innovation Skills; Information, Media and Technology Skills; Life and Career Skills

Table 4.2. Key Vocabulary in Lesson 1

Key Vocabulary	Definition
attract	to pull toward
cartography	the interpretation of geographic information to create maps and charts
climate map	a map that shows an area's typical weather patterns
destination	a place to which people go
economic map	a map that shows the primary types of resources or economic activities in an area; also known as a *resource map*
engineer	a person who designs and builds products to meet human needs
force	a push or pull between objects
geography	the study of the earth's natural land formations and human society, with a focus on the relationship between physical features of the earth and human activity
levitation	the process of the rising or lifting of an object with no visible means of support
maglev	a method of transportation using magnetic levitation, with vehicles suspended above the tracks by magnetism
magnetic field	an invisible area of magnetism around a magnet
magnetic poles	the parts of the magnet that exhibit magnetic properties, classified as either north or south
magnetism	the force of attraction and repulsion between two magnetic materials
map scale	the relationship between the distance on a map and the actual distance
physical map	a map that shows an area's physical features, such as mountains, lakes, and canyons
political map	a map that shows states' and countries' boundaries, along with major cities
region	an area of a country or the world that has a set of distinct characteristics
repel	to push away
road map	a map that shows an area's roads, airports, railroad tracks, and points of interest
topographic map	a map that includes contour lines to show an area's physical characteristics, such as hills and valleys

TEACHER BACKGROUND INFORMATION
Engineering

Engineering is a topic with which many teachers and students have little experience. Third-grade students may find references to engineers confusing in this module because they may have heard the term used in reference to the railroad engineers who operate trains. In this unit, the term *engineer* is used to denote individuals in the railway engineering profession, which falls under the broad umbrella of transportation engineering. Students should understand that trains are designed by engineers trained in areas such as mechanical engineering, civil engineering, electrical engineering, computer engineering, and even aerospace engineering for high-speed trains. For an overview of the various types of engineering professions, see the following websites:

- *www.nacme.org/types-of-engineering*

- *www.engineeryourlife.org/?ID=6168*

- *www.sciencekids.co.nz/sciencefacts/engineering/typesofengineeringjobs.html*

Geography and Map Skills

Students learn about the basic geography of the United States in this module. They will become familiar with the five regions of the United States—the Northeast, Southwest, West, Southeast, and Midwest—and understand that regions are areas of land grouped according to their location that may have common natural and cultural features. A map of U.S. regions can be found at *http://media.education.nationalgeographic.com/assets/file/us-regions-map.pdf*.

There are various types of maps. Students may be familiar with road maps and online mapping websites such as Google Maps. Other types of maps include topographic maps, relief maps, political maps, and weather maps. Road maps are the primary map resource used in this module, although topographic maps may be used as well. The U.S. Geological Survey (USGS) provides a number of educational resources for working with maps. These excellent aids for supporting your students' basic map-reading skills can be found at *http://education.usgs.gov/primary.html#geoggeneral*.

Career Connections

As career connections related to this lesson, you may wish to introduce the following:

- *Geographer:* Geographers study the Earth's natural land formations and human society, with a focus on the relationship between these phenomena. In particular, they study the characteristics of various parts of the Earth, including physical characteristics and human culture. Many geographers work for the federal government. Teaching and field research are other areas in which geographers work. For more information, see *www.bls.gov/ooh/life-physical-and-social-science/geographers.htm*.

- *Cartographer:* Cartography is a subset of geography. Cartographers interpret geographic information and create maps and charts. Many cartographers have backgrounds in geography and civil engineering. For more information, see *www.bls.gov/ooh/architecture-and-engineering/cartographers-and-photogrammetrists.htm*.

- *Photogrammetrist:* Photogrammetrists use aerial photographs, satellite imagery, and other images to create maps or drawings of geographic areas. This field is closely related to cartography. For more information, see the above website as well as *www.wisegeek.com/what-are-photogrammetrists.htm*.

Trains

The focus in this lesson is on train travel, and teachers should access students' prior experiences and current understanding and perceptions. Students will have had varying experiences with train travel. This may depend on where they live, since passenger train travel is much more prevalent in the Northeast than in some other parts of the country such as the Midwest, where freight trains are more common. Showing the class video footage of modern train technology may influence students' perceptions.

This module focuses more specifically on maglev train technology. *Maglev* is a term that combines the words *magnetic* and *levitation* and is used to refer to modes of transportation in which the vehicle (typically a train) travels without touching the ground, using magnets to provide lift and forward movement. For more information about maglev train technology and a proposal to provide a maglev train between New York and Washington, D.C., see the following websites and YouTube video:

- *http://science.howstuffworks.com/transport/engines-equipment/maglev-train.htm*

- *www.wsj.com/articles/campaign-for-floating-train-to-connect-new-york-and-washington-gathers-pace-1413976169*

- *www.youtube.com/watch?v=aIwbrZ4knpg*

Magnets and Magnetism

Most third-grade students have had hands-on experience with magnets and thus some understanding of magnetism, but they may also have some conceptual misunderstandings about how magnets work. Teachers should be ready to provide additional content information about magnetism throughout the unit.

Magnetism, as the term is used in this module, can be defined as the force of attraction or repulsion of materials, typically metals such as iron, steel, and nickel. More simply, it refers to the ability of objects to attract iron. The alignment of electrically charged particles within a substance causes magnetism. Magnetism can be permanent or temporary. Permanent magnets are those in which the magnetism is a physical characteristic of the substance. Temporary magnetism results when an object is moved by a permanent magnet and typically disappears when the permanent magnet is removed.

Magnetic objects create a magnetic field, which is the space around the magnet in which the magnetic force exists. Opposing magnetic forces are produced at each end, or pole, of a magnet. When a magnet is suspended (with no friction), it will automatically orient itself so that one pole points north and one south, and thus, the ends are labeled the north and south poles. If you work with multiple magnets, opposite poles will attract each other, and like poles will repel each other. For instance, if you lay two magnets on a table so that opposite ends are together, they will be drawn together, but if you place them so that the two south poles (or north poles) touch, they will separate. A good way for students to remember this is to use the phrase "opposites attract." The poles of a magnet are not visually distinguishable as either north or south; however, if you place a magnet beside a compass, the needle that points toward the Earth's north will move toward the magnet's south pole, allowing you to distinguish between north and south poles.

For information about magnets and compasses, see the following websites:

- *http://science.howstuffworks.com/magnetism-channel.htm*

- *www.livescience.com/38059-magnetism.html*

- *www.nde-ed.org/EducationResources/HighSchool/Magnetism/twoends.htm*

COMMON MISCONCEPTIONS

Students will have various types of prior knowledge about the concepts introduced in this lesson. Table 4.3 (p. 52) outlines some common misconceptions students may have concerning these concepts. Because of the breadth of students' experiences, it is not possible to anticipate every misconception that students may bring as they approach this lesson. Incorrect or inaccurate prior understanding of concepts can influence student learning in the future, however, so it is important to be alert to misconceptions such as those presented in the table.

Table 4.3. Common Misconceptions About the Concepts in Lesson 1

Topic	Student Misconception	Explanation
Magnetism	All metals are magnetic.	The only naturally occurring magnetic metals are iron, cobalt, and nickel.
	All magnets are solids.	Magnetic fields can be created in space by electric currents.
	Large magnets exert a stronger magnetic field than small magnets.	The size of a magnet and its magnetism are not necessarily related. The substances that compose the magnet, not its size, determine its magnetic force.
Maps*	Map symbols directly represent their referents.	Symbols may be abstract and may not be a true physical representation of the feature. For example, blue on a map does not always indicate water, a triangle does not always indicate a mountain, and green areas do not always indicate trees. The map legend provides a key for symbols.

*For more information on misconceptions students might have about maps, see the National Geographic report Spatial Thinking About Maps: Development of Concepts and Skills Across the Early Years at *http://nationalgeographic.org/media/spatial-thinking-about-maps.*

PREPARATION FOR LESSON 1

Review the Teacher Background Information, assemble the materials for the lesson, and preview the videos recommended in the Learning Plan Components section below. Prepare plastic sandwich bags with iron filings and index cards inside for the Magnet Magic activity, and seal the bags securely with duct tape. Students should not open these during the activity, as iron filings can be hazardous if they get into the eyes or are inhaled or swallowed. Consult the safety data sheet that accompanies the iron filings for full information. Be sure to wear safety glasses and non-latex gloves at all times when handling filings. Use caution when handling sharp objects such as tacks or nails, as they can cut or puncture skin. Wash hands with soap and water after handling the filings and completing the activity.

Have your students set up their STEM Research Notebooks (see pp. 25–26 for discussion and student instruction handout).

LEARNING PLAN COMPONENTS
Introductory Activity/Engagement

Connection to the Challenge: Begin each day of this lesson by directing students' attention to the driving question for the module and challenge: How can we create a plan and build a prototype for a maglev train to carry passengers to a vacation destination? Hold a brief student discussion of how their learning in the previous days' lessons contributed to their ability to create their plan and build their prototype. You may wish to hold a class discussion, creating a class list of key ideas on chart paper, or you might have students create a notebook entry with this information.

Social Studies Class: Social studies is one of the driving content areas behind this interdisciplinary unit of study and therefore may be used as an opportunity to launch the module. Begin the lesson by asking students how many of them have ridden on a train. Find out what students know and think about train travel by asking them questions such as the following and recording their responses (you may wish to create a KWL chart for this):

- Did you ever see a train?

- What do you think the train was carrying?

- Have you ever ridden on a train? (allow students to share personal experiences)

- Do you think that many people in the United States travel on trains?

- How fast do you think trains can go?

- Do you think that trains could go faster than they do now?

- Where can you go on a train?

- How are trains powered? Do they use gas like cars, electricity, or something else?

- Can you board a passenger train in our town?

Introduce modern trains by showing a video about futuristic transportation. (Relevant videos can be found on YouTube by searching for "future of transportation technologies"; one example is "Hyperloop and Future Transport Technology" at *www.youtube.com/watch?v=YHiKjJEFY6A*). After watching the video, ask students what they learned about trains. What were they surprised about? Can they identify some of the ways trains have changed over the years? Enter this information in the KWL chart.

Tell students that they will be working as train design engineers in this module. They will be challenged to create a Maglevacation Train. Use the Maglevacation Train Challenge graphic (p. 157) to introduce students to the challenge. Hold a class brainstorming session to create a list of what students need to know to complete this challenge (e.g., how a maglev train works, where the train needs to go).

Tell students that they will be learning about trains and mapping a route in the coming weeks, but first they need to form their design teams. Divide students into groups of three to four; these will be the design teams for the rest of the module. Each group of students should complete three tasks:

1. Choose a name for their design team.

2. Brainstorm a list of at least three possible vacation destinations. (Tell teams that the destinations must be within the continental United States and more than 100 miles from their hometown.)

3. Agree on the top three or four destinations and create a list (number of destinations should match number of team members).

Have each team member take responsibility for one of the destinations and gather information about that place (this may be assigned as homework), including the following:

- Location (city or town and state)

- Major attractions (e.g., amusement parks, beaches)

- Information about the climate and weather (e.g., average temperatures, whether it is warm year-round)

Science Class: Tell students that in order to create a maglev train, they first need to understand magnets and magnetism. To assess students' current understanding of magnets and magnetism and identify misconceptions that need to be addressed, have students respond to the following preassessment questions in their STEM Research Notebooks:

- Show students a magnet and five or six objects (see suggested items in the list of Additional Materials for Magnetic or Not? activity) and ask about each, Can I pick up this item with the magnet?

- How can you tell if an item will be attracted by a magnet?

- How can magnets be used in everyday life?

Then ask students to share what they know about magnets and what they want to know about them (guide students to relate their questions to maglev trains and the role of magnetism in moving the train). Create a KWL chart from this discussion.

Introduce the idea of properties of magnets by telling students that everything they see (all matter) has what we call physical properties, which describe how the object looks, feels, and acts. We can observe and measure these properties. Show students a ball. Ask students to share what properties they can observe about the ball. Hold the ball over a nail; ask students what they observe. Next, show them the magnet; ask students to share what properties they can observe about the magnet. Now, hold the magnet over the nail; ask students what they observe. Show students a variety of magnets and ask them what properties these magnets share. Create a class list of the properties of magnets.

Mathematics Connection: Not applicable.

ELA Connection: Have students design covers for their STEM Research Notebooks that reflect the interdisciplinary nature of this unit.

Activity/Exploration

Social Studies Class: Students explore the geography of the United States using maps and identify the region, state, county, and town or city in which they live in the Map Me! activity.

Map Me!

Show the class a map of the regions of the United States (an example can be found at *http://media.education.nationalgeographic.com/assets/file/us-regions-map.pdf*). Ask students to identify what region they live in and share what they believe are some features of this region that make it unique and distinguish it from other regions. Have students create a Research Notebook entry describing the geographic features of the city or town in which they live as well as the region. Review map-reading skills with students (see Explanation section on p. 58 for more detail), and demonstrate measuring distances on a map using the map scale and a ruler. Have students work in their design teams to complete the Map Me! handouts (attached at the end of this lesson).

After groups have completed the handouts, tell them that they must choose a final destination for their team. Remind them that this is the destination to which their maglev trains will be traveling. Point out to students that if they draw a straight line from their hometown to the destination, it will be shorter than the route on the road map. Have a class discussion about why this is so (e.g., roads are usually not completely straight, as they must avoid geographic barriers such as mountains, canyons, and large bodies of water and must go to multiple destinations). Demonstrate to students how to measure road distances on a road map using a ruler. If roads are curved, show students how to lay a piece of string on the map following the road, mark the starting point and end point on the string, and then measure between those points on the string to determine distance. Once they have decided on a destination, have students mark the location (town or city and state) on their blank maps and record the distance, traveling by road, between their hometown and the destination on this map.

Science Class: Introduce the basic concept of magnetism to students (see Explanation section on p. 58 for more details). Students will investigate magnetism through two inquiry activities, Magnetic or Not? and Magnet Magic.

Magnetic or Not?

Students will explore magnetism using a "fishing pole" made from a pencil, string, and magnet. Group students in teams of two to three, and have each team tie a magnet onto one end of an 18-inch piece of string, then tie the other end of the string to a pencil to create the "fishing pole." Ask students to predict whether the magnet will pick up each kind of material. Have students record their predictions on the Magnetic or Not? handout (attached at the end of this lesson). Next, give students time to test their magnets with the objects provided and record their observations on the Magnetic or Not? handout. Ask students if they observe any similarities between the materials that were magnetic and how they are different from those that were not. Have students record their conclusions about the similarities and differences on the Magnetic or Not? handout. Refer to the list of properties of magnets the class created in the Introductory Activity/Engagement section on page 54, and have students compare and contrast their conclusions from this activity with the properties on the list. Add to the class list as necessary.

Next, separate out the materials that have magnetic properties, and ask students how close they think the magnet needs to be to each magnetic material to pick it up. Show students how to hold a ruler vertically beside the magnet as you lower it toward the object to determine the distance. Have students predict the distance for each magnetic object, and then test those predictions. Students should record their predictions and observations on the Magnetic or Not? handouts. Use this information to introduce the notion of a magnetic field (for more information, see the Teacher Background Information on magnets and Explanation sections on pp. 50 and 58). Have students include the Magnetic or Not? handout in their Research Notebooks.

Show students the video "Magnet Mania!" (visit YouTube and search for "Magnet Mania!" or access the video directly at *www.youtube.com/watch?v=2QiyiWLm2FY*). Have students review the list of uses of magnets from the preassessment before watching the video, and create a class list of their ideas. Ask students to watch for other uses of magnets in the video. After viewing the video, add to the class list of uses of magnets.

Tell students that magnets interact not only with other substances but also with each other. In the Magnet Magic activity, students will learn that the way magnets interact with each other depends on which poles of the magnets are close to each other.

Magnet Magic

Students should work in pairs for this activity. Instructions are provided on the Magnet Magic handouts. Iron filings can be hazardous if they get into the eyes or are inhaled or

swallowed. Consult the safety data that accompanies the iron filings for full information. Be sure to have students wear safety glasses and gloves at all times during this activity. The iron filings should be securely sealed in plastic sandwich bags along with the index cards (see Preparation for Lesson 1, p. 52) before giving the bags to students. Be sure that students do not open or tear the bags, blow on the filings, or allow the filings to come into contact with their skin. Remember to wash hands with soap and water after completing this activity.

After the activity, hold a class discussion on what students observed about magnetic fields and how they could use magnets for motion. Have students create an entry in their STEM Research Notebooks to reflect on how they could use what they learned in this activity to create their maglev trains.

Mathematics Connection: Assign various routes along a map and ask students to calculate the distances between points. Have students include these calculations as a Research Notebook entry. Introduce the concept of speed (distance divided by time), and discuss with students how speed affects travel time (for more details, see Mathematics Connection in the Explanation section, p. 58). Discuss the concept of speed limits and how speeds vary when traveling.

ELA Connection: Students should create a table of contents page for their Research Notebooks. You may wish to use this task as an opportunity to distinguish between indexes and tables of contents. You might also want to launch a discussion about the differences and similarities between fiction and nonfiction texts (see ELA Connection in the Explanation section on p. 59 for more information). Students can compare and contrast fiction and nonfiction texts through a classroom discussion and by creating a Venn diagram. A variety of fiction and nonfiction texts relevant to the lesson topics may be explored throughout the module. You may wish to have students create Research Notebook entries in each lesson to reflect on the readings and how they may connect to the Maglevacation Train Challenge. Suggested literature connections include the following:

- *Amazing Magnetism,* by Rebecca Carmi (Scholastic; ISBN: 978-0439314329)

- *Locomotive,* by Brian Floca (Atheneum; ISBN: 978-1416994152)

- *Magnets: Pulling Together, Pushing Apart,* by Natalie M. Rosinsky (Turtleback; ISBN: 978-1404803336)

- *Mapping Penny's World,* by Loreen Leedy (Square Fish; ISBN: 978-0805072624)

- *My America: A Poetry Atlas of the United States,* by Lee Bennett Hopkins (Simon & Schuster; ISBN: 978-0689812477)

- *The Boxcar Children,* by Gertrude Chandler Warner (Albert Whitman & Company; ISBN: 978-0807508527)

- *Trains*, by Mary Lindeen (Bellwether Media; ISBN: 978-1600140624)

- *What Makes a Magnet?*, by Franklyn M. Branley (HarperCollins; ISBN: 978-0064451482)

Explanation

This section contains an overview of concepts students should understand in this lesson. You may also wish to have students share what they have learned and explain their understanding of lesson concepts.

Social Studies Class: Students will learn that prototypes are models of a product used for testing a product. These are often at a small scale and are used for repeated testing so that improvements can be made. Explain to students that there are many different types of maps, including road maps, topographic maps, and relief maps, and that each serves a different purpose. The USGS provides downloadable samples of these different kinds of maps that you may wish to show your students. See *http://egsc.usgs.gov/isb//pubs/teachers-packets/mapshow/posterandpacket.html* for these resources.

For this module, road maps will be the primary map resource, although topographic maps may be used as well. The USGS provides educational materials about reading and using maps, and you may wish to access these resources to support students' map-reading skills. These may be accessed at *http://education.usgs.gov/primary.html#geoggeneral*.

Introduce careers associated with mapping. These include geographers, cartographers, and photogrammetrists (see Teacher Background Information section on p. 49 for more information).

Science Class: Introduce the concept of magnetism to students as a force that causes a substance to attract or repel other objects (for more details, see the Teacher Background Information section on p. 50). Students should understand the following points for this module:

- Opposite poles of magnets attract; like poles repel.

- Magnets are surrounded by an invisible force called a magnetic field.

- The magnetic forces of attraction or repulsion can vary depending on an object's location within the magnetic field (i.e., if an object is closer to the magnet, it is more likely to be attracted).

- Magnetic fields are not affected by the presence of nonmagnetic materials.

- Magnets can be used to cause objects to move.

Mathematics Connection: Students will need to understand the concept of distances for this module and how to calculate those distances using a map. Explain to students that speed is the distance that is traveled divided by the time it has taken to travel that distance. The equation Speed = Distance/Time is used to calculate speed. You may wish to use speed limit signs as an example and note to students that instead of saying "divided by," we say "per" when talking about speed limits (as in "miles per hour") to reflect the distance we could travel in one hour at that speed. Emphasize to students that speed limits are based on an hour's travel time. Discuss the units used for speed, emphasizing that two units—miles and hours—must be given. Model how to use this equation. For instance, if we travel 40 miles in 2 hours, we calculate our speed as follows: Speed = 40 miles/2 hours = 20 miles per hour.

ELA Connection: In third grade, students begin to explore increasing numbers of nonfiction texts in a variety of disciplines and content areas. Students may experience the concepts in this lesson through reading a variety of nonfiction and fiction texts, although the majority of the literature connections provided for this lesson are nonfiction. This provides an opportunity to allow students to investigate the characteristics of fiction versus nonfiction texts. Table 4.4 lists some points that you may wish to emphasize. Students may add more items as they brainstorm.

Table 4.4. Characteristics of Nonfiction and Fiction Texts

Nonfiction	Fiction
Conveys facts and information	Tells a story
Is read to learn	Is read to enjoy
May include photos, charts, and graphs	May include illustrations
Presents information and directions	Has characters, a setting, and a plot
May have a table of contents, glossary, and index	May have separate chapters

Elaboration/Application of Knowledge

Social Studies Class: Students can apply their growing understanding of maps and map reading in numerous ways. The following are suggestions for extension activities: Have students locate the major rivers in the United States, including the Missouri, Mississippi, Colorado, and Ohio Rivers, and mark them on their map templates from the Map Me! student handouts. Discuss with students what these rivers mean for railroads (i.e., the need for bridges). Then, have students do the same for the major mountain ranges in the United States and discuss how railroads navigate mountains (e.g., via tunnels or bridges, or by going around them). See *http://gsrj.com/greatsmokypremium.html* for a video tour of the Great Smoky Mountains Railroad featuring a number of tunnels and bridges.

Students may explore a variety of mapping and geography skills at the following websites:

- Map reading and mileage calculations (Harcourt educational website): *www.harcourtschool.com/activity/road_maps/index.html*

- Train traffic control (interactive game): *http://alltraingames.com/trains/train-traffic-control-.html*

Then, have students create a STEM Research Notebook entry in response to the prompt below.

STEM Research Notebook Prompt

Students should respond to the following prompt in their Research Notebooks: *Many people do not use paper maps when they travel today. Instead, they use electronic tools such as Global Positioning System (GPS) devices in their cars or on their cell phones. Describe some advantages and disadvantages of using maps and of using GPS devices. How is the information they provide to travelers different? How is the information the same?*

Science Class: Assign students a magnetic scavenger hunt as a homework assignment. Give each student a bar magnet to take home and ask them to identify magnetic materials in their house. (*Safety note:* Make sure students and parents have signed Safety Acknowledgement Forms relative to standard safety operating procedures beforehand.) Students can classify materials as "Strong Attraction," "Weak Attraction," and "No Attraction" and create a chart of their findings. Be sure to tell students not to place their magnets near cell phones, computers, or cards with magnetized strips, such as credit cards.

Mathematics Connection: Students should research the cost of a passenger train ticket from their hometown to the vacation destination their group chose. Then have students compare this with the cost of an airline ticket, a bus ticket, or other modes of transportation. Have students use division to calculate the cost per mile of each mode of transportation and then create a bar graph comparing these costs.

ELA Connection: Have students create a Reading Response section for their STEM Research Notebooks. You may choose to use a variety of prompts based on the selected readings. This can be continued throughout the module with various pieces of literature.

Nonfiction examples include the following:

- What is the most important thing the author wants you to understand or know after reading this book? How do you know that?

- What surprised you the most in this book?

- How can what you learned in this book help you in the Maglevacation Train Challenge?

- What kinds of text features did this book use to help you learn? How did these affect your reading experience?

- Did this book remind you of another book, a movie, a real-life experience, or a current event? Why?

- What would you like to know more about after reading this book? Why? How could you find out more about this?

Fiction examples include the following:

- Would you have liked to live in the time and place in which this book was set? Why or why not?

- What character in this book would you choose to have as a friend? Why?

- What real-life events does this book remind you of? Why?

- Why do you think the author chose to write this story?

- From what you have read so far, what do you think will happen next?

Evaluation/Assessment

Students may be assessed on the following performance tasks and other measures listed.

Performance Tasks

- Map Me! handouts

- Map Me! map activity

- Magnetic or Not? handout

- Magnet Magic handouts

- Literature reading response Research Notebook entry

Other Measures

- Other STEM Research Notebook entries

- Participation in class and group discussions

INTERNET RESOURCES

National Geographic "Spatial Thinking About Maps: Development of Concepts and Skills Across the Early Years"

- *http://nationalgeographic.org/media/spatial-thinking-about-maps*

"Hyperloop and Future Transport Technology" video

- *www.youtube.com/watch?v=YHiKjJEFY6A*

Map of U.S. regions

- *http://media.education.nationalgeographic.com/assets/file/us-regions-map.pdf*

USGS educational map-reading resources

- *http://education.usgs.gov/primary.html#geoggeneral*

Information about careers

- *www.bls.gov/ooh/life-physical-and-social-science/geographers.htm*
- *www.bls.gov/ooh/architecture-and-engineering/cartographers-and-photogrammetrists.htm*
- *www.wisegeek.com/what-are-photogrammetrists.htm*

Information about maglev trains

- *http://science.howstuffworks.com/transport/engines-equipment/maglev-train.htm*
- *www.wsj.com/articles/campaign-for-floating-train-to-connect-new-york-and-washington-gathers-pace-1413976169*
- *www.youtube.com/watch?v=aIwbrZ4knpg*

Information about magnets

- *http://science.howstuffworks.com/magnetism-channel.htm*
- *www.livescience.com/38059-magnetism.html*

Downloadable samples of types of maps

- *http://egsc.usgs.gov/isb//pubs/teachers-packets/mapshow/posterandpacket.html*

Video tour of the Great Smoky Mountains Railroad

- *http://gsrj.com/greatsmokypremium.html*

Map reading and mileage calculations for student exploration

- *http://www.harcourtschool.com/activity/road_maps/index.html*

Interactive game on the train traffic control

- *http://alltraingames.com/trains/train-traffic-control-.html*

STUDENT HANDOUT, PAGE 1

MAP ME!

Name: _____

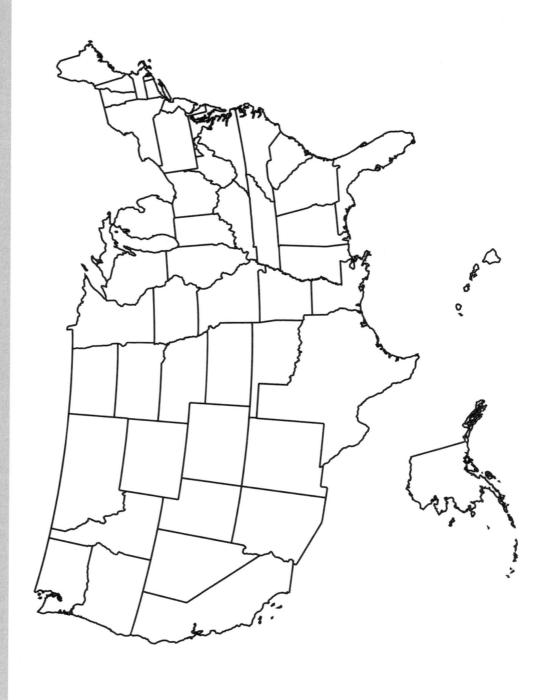

Name: _____

STUDENT HANDOUT, PAGE 2

MAP ME!

1. With your group, do the following:

Use the road map to find the town or city where you live, and fill in the following information:

I live in _____ (city or town), in

_____ (county), in

_____ (state).

My state is in the _____ region of the United States.

2. Mark the location of your town on your blank map, and label your state, county, and town or city.

3. Look at your group's list of possible vacation destinations. If you do not know the state and town for these, use the internet to find this information.

4. Find each of the vacation destinations on the road map, and circle it with pencil.

5. Use the map scale and your ruler find out how far from your hometown each of the possible vacation destinations is, and record that information:

Possibility 1:

Name of destination: _____

State: _____

City or town: _____

Distance (in miles) from home: _____

Name: _____

STUDENT HANDOUT, PAGE 3

MAP ME!

Possibility 2:

Name of destination: _____

State: _____

City or town: _____

Distance (in miles) from home: _____

Possibility 3:

Name of destination: _____

State: _____

City or town: _____

Distance (in miles) from home: _____

4

Name: _____

STUDENT HANDOUT

MAGNETIC OR NOT?

OBJECT	PREDICT Will it be magnetic?	OBSERVE Was it magnetic?	PREDICT How many inches away does the magnet need to be to pick the object up?	OBSERVE How many inches away was the magnet when it picked the object up?

What similarities do you notice between materials that were magnetic? _____

_____

How are these materials different from the materials that were not magnetic? _____

_____

Name: _____

STUDENT HANDOUT, PAGE 1

MAGNET MAGIC, PART 1

There are three parts to this activity. Follow the directions for each and record your observations.

PART 1

Safety note: Always keep your sandwich bag sealed and intact during this activity! Also be sure to have on eye protection (safety glasses or goggles) and gloves.

1. Lay the plastic sandwich bag with the iron filings and index card inside it flat on the table.

2. Shake the bag gently so that there is a thin layer of iron filings on top of the index card. Draw what you see:

3. Lift the bag carefully so that you don't disturb the filings, place a paper clip on the table, and then put the bag on top of it. What happened?

   _____

   _____

4. What will happen if you put a magnet underneath the bag instead? Make a prediction:

   _____

   _____

4

Name: _____

STUDENT HANDOUT, PAGE 2

MAGNET MAGIC, PART 1

5. Lift the bag carefully so that you don't disturb the filings, remove the paper clip, and put a bar magnet underneath the bag. What happened? Draw what you see:

6. Why do you think this happened? _____

   _____

7. Every magnet has an invisible field around it called a magnetic field. Your drawing shows you the magnetic field. It is invisible, but the iron filings show where it is because they line up with it. Label the magnetic field on your drawing.

8. Now, put the bag aside and get your second bar magnet. Remember that magnets have two poles—north and south—and that opposite poles attract. Find each magnet's poles by putting the magnets end to end and seeing which ends attract and repel.

9. Carefully shake the iron filings so they are distributed evenly over your card again. Place your two bar magnets on the table so that the ends are repelling each other. Put the sandwich bag on top and draw what you see:

Name: _____

STUDENT HANDOUT, PAGE 3

MAGNET MAGIC, PART 1

10. Try this with some other combinations of magnets, and draw the magnetic fields you see:

11. How did the magnetic fields change with the different combinations of magnets?

_____

_____

_____

_____

_____

_____

_____

Name: _____

STUDENT HANDOUT

MAGNET MAGIC, PARTS 2 AND 3

PART 2

Safety note: Be sure to immediately wipe up any spilled water, which could cause someone to slip and fall!

1. Fill your cup with water to about 2 inches below the top.

2. Now, drop a round magnet into the cup and let it sink to the bottom.

3. Your challenge is to remove the magnet from the cup without getting your fingers or any other object wet. You may use any of your other magnets.

4. How did you get your magnet out of the cup?

_____

_____

PART 3

1. Lay the Magnet Magic Maze handout on a flat surface.

2. Place one of the small, round magnets at the start of the maze.

3. Using another of your magnets, move the magnet through the maze. You may not touch the magnet in the maze with your hands or any other object.

4. How did you get your magnet out of the maze?

_____

_____

REFLECTION

Use this section to describe what you learned about magnets and magnetic fields in this activity:

_____

_____

Name: _____

STUDENT HANDOUT

EXPLORATION

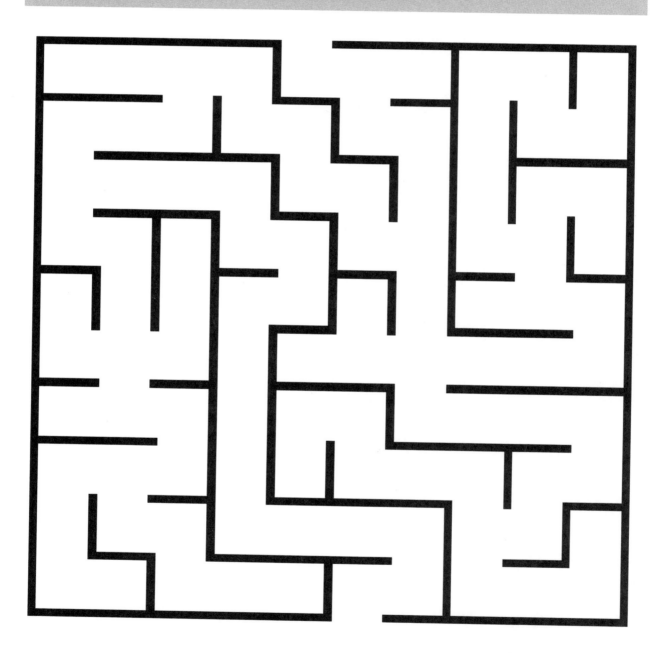

Lesson Plan 2: Trains Through Time

This lesson provides information to enable students to identify the connection between the westward expansion of the United States and the proliferation of train travel. The social studies component of this lesson includes the history of trains as it relates to the expansion of the United States, with a focus on the Transcontinental Railroad. Students are also introduced to information about the railway engineers who are responsible for train design. Students extend and apply this learning as they gain an understanding of passenger rail travel in the 21st century and its availability in various regions across the country. They formulate research questions and design their own magnet investigations. Students are introduced to the EDP in this lesson and use this process to structure their participation in a mini design challenge, the Magnificent Magnet Match. In mathematics, students use railroad timetables to explore elapsed time and extend this understanding as they learn about time zones and their connection to railroads. In ELA, students make a literature connection to what they are learning as they explore train engineering and biomimicry using the book *From Kingfishers to … Bullet Trains*.

ESSENTIAL QUESTIONS

- How did train travel influence the geographic growth of the United States?
- What is the prevalence of passenger train travel in various regions across the United States?
- What is biomimicry and how does it apply to train travel?
- How can we estimate the distance between two cities?
- How can we design an investigation that will provide us with information about magnets?
- How do engineers structure their work in designing products and solutions to problems?

ESTABLISHED GOALS AND OBJECTIVES

At the conclusion of this lesson, students will be able to do the following:

- Provide examples of how trains contributed to the geographic expansion of the United States
- Identify the span of the Transcontinental Railroad on a map and understand how this correlated to westward expansion in the United States

- Place the history of trains in the United States within the international development of trains beginning with wagon tramways in the 1500s

- Provide examples of biomimicry in train technology

- Estimate and determine the actual mileage between two cities

- Understand the concept of elapsed time and calculate travel times between various time zones

- Use the EDP to work collaboratively

- Apply their understanding of testable questions to formulate a testable research question related to magnets

- Design an investigation related to magnets

- Apply their understanding of magnetism to a group design challenge

- Demonstrate their understanding of the EDP

TIME REQUIRED

- 4 days (approximately 45 minutes each day; see Tables 3.6–3.7)

MATERIALS

Required Materials for Lesson 2

- STEM Research Notebooks

- Access to internet technology for showing video clips and for student research

- Classroom map of the United States

- Handouts (attached at the end of this lesson)

- Time zone map and EDP graphic to share with class (attached at the end of this lesson)

- Train timetables (current Amtrak timetables can be found at *www.amtrak.com/ train-schedules-timetables*)

- *If You Traveled West in a Covered Wagon,* by Ellen Levine (Scholastic; ISBN: 978-0590451581)

Additional Materials for Riding the Rails

- Masking tape or clothesline and clips (to hang timeline)

- Map Me! handouts (1 set per student)

Additional Materials for My Magnetic Question

- Materials as determined by student plans (e.g., magnets, paper, water, paper clips)

Additional Materials for Magnificent Magnet Match (per team unless otherwise indicated)

- Safety glasses or goggles (1 pair per student)
- Bar of soap
- 2 small, round magnets
- 2 bar magnets
- 3 rubber bands
- 2 toothpicks
- Construction paper
- 1 roll of masking tape
- EDP graphic
- Engineer It! handouts (1 set per student)
- Wading pool filled with water (1 per class)

CONTENT STANDARDS AND KEY VOCABULARY

Table 4.5 (p. 76) lists the content standards from the *NGSS*, *CCSS*, and Framework for 21st Century Learning that this lesson addresses, and Table 4.6 (p. 80) presents the key vocabulary. Vocabulary terms are provided for both teacher and student use. Teachers may choose to introduce some or all of the terms to students.

Table 4.5. Content Standards Addressed in STEM Road Map Module Lesson 2

NEXT GENERATION SCIENCE STANDARDS

PERFORMANCE EXPECTATIONS

- 3-PS2-3. Ask questions to determine cause and effect relationships of electric or magnetic interactions between two objects not in contact with each other.

- 3-PS2-4. Define a simple design problem that can be solved by applying scientific ideas about magnets.

- 3-5-ETS1-1. Define a simple design problem reflecting a need or a want that includes specified criteria for success and constraints on materials, time, or cost.

- 3-5-ETS1-2. Generate and compare multiple possible solutions to a problem based on how well each is likely to meet the criteria and constraints of the problem.

- 3-5-ETS1-3. Plan and carry out fair tests in which variables are controlled and failure points are considered to identify aspects of a model or prototype that can be improved.

SCIENCE AND ENGINEERING PRACTICES

Asking Questions and Defining Problems

- Ask questions about what would happen if a variable is changed.

- Use prior knowledge to describe problems that can be solved.

- Define a simple design problem that can be solved through the development of an object, tool, process, or system and includes several criteria for success and constraints on materials, time, or cost.

Developing and Using Models

- Identify limitations of models.

- Collaboratively develop and/or revise a model based on evidence that shows the relationships among variables for frequent and regular occurring events.

- Develop a diagram or simple physical prototype to convey a proposed object, tool, or process.

- Use a model to test cause and effect relationships or interactions concerning the functioning of a natural or designed system.

Planning and Carrying Out Investigations

- Plan and conduct an investigation collaboratively to produce data to serve as the basis for evidence, using fair tests in which variables are controlled and the number of trials considered.

- Make observations and/or measurements to produce data to serve as the basis for evidence for an explanation of a phenomenon or test a design solution.

Analyzing and Interpreting Data
- Analyze and interpret data to make sense of phenomena, using logical reasoning, mathematics, and/or computation.
- Compare and contrast data collected by different groups in order to discuss similarities and differences in their findings.
- Analyze data to refine a problem statement or the design of a proposed object, tool, or process.
- Use data to evaluate and refine design solutions.

Using Mathematical and Computational Thinking
- Organize simple data sets to reveal patterns that suggest relationships.

Constructing Explanations and Designing Solutions
- Apply scientific ideas to solve design problems.
- Generate and compare multiple solutions to a problem based on how well they meet the criteria and constraints of the design solution.

DISCIPLINARY CORE IDEAS

PS2.A: Forces and Motion
- Each force acts on one particular object and has both strength and a direction. An object at rest typically has multiple forces acting on it, but they add to give zero net force on the object. Forces that do not sum to zero can cause changes in the object's speed or direction of motion.
- The patterns of an object's motion in various situations can be observed and measured; when that past motion exhibits a regular pattern, future motion can be predicted from it.

PS2.B: Types of Interactions
- Objects in contact exert forces on each other.
- Electric and magnetic forces between a pair of objects do not require that the objects be in contact. The sizes of the forces in each situation depend on the properties of the objects and their distances apart and, for forces between two magnets, on their orientation relative to each other.

ETS1.A: Defining and Delimiting Engineering Problems
- Possible solutions to a problem are limited by available materials and resources (constraints). The success of a designed solution is determined by considering the desired features of a solution (criteria). Different proposals for solutions can be compared on the basis of how well each one meets the specified criteria for success or how well each takes the constraints into account. (3-5-ETS1-1)

ETS1.B: Developing Possible Solutions
- Research on a problem should be carried out before beginning to design a solution. Testing a solution involves investigating how well it performs under a range of likely conditions. (3-5-ETS1-2)
- At whatever stage, communicating with peers about proposed solutions is an important part of the design process, and shared ideas can lead to improved designs. (3-5-ETS1-2)
- Tests are often designed to identify failure points or difficulties, which suggest the elements of the design that need to be improved. (3-5-ETS1-3)

ETS1.C: Optimizing the Design Solution
- Different solutions need to be tested in order to determine which of them best solves the problem, given the criteria and the constraints. (3-5-ETS1-3)

CROSSCUTTING CONCEPTS

Patterns
- Similarities and differences in patterns can be used to sort, classify, communicate, and analyze simple rates of change for natural phenomena and designed products.
- Patterns of change can be used to make predictions.
- Patterns can be used as evidence to support an explanation

Scale, Proportion, and Quantity
- Standard units are used to measure and describe physical quantities such as weight, time, temperature, and volume.

Energy and Matter
- Energy can be transferred in various ways and between objects.

Cause and Effect
- Cause and effect relationships are routinely identified, tested, and used to explain change.

Stability and Change
- Change is measured in terms of differences over time and may occur at different rates.

Influence of Science, Engineering, and Technology on Society and the Natural World
- People's needs and wants change over time, as do their demands for new and improved technologies.
- Engineers improve existing technologies or develop new ones to increase their benefits, decrease known risks, and meet societal demands.

COMMON CORE STATE STANDARDS FOR MATHEMATICS

MATHEMATICAL PRACTICES

- MP1. Make sense of problems and persevere in solving them.
- MP2. Reason abstractly and quantitatively.
- MP4. Model with mathematics.
- MP5. Use appropriate tools strategically.
- MP6. Attend to precision.

MATHEMATICAL CONTENT

- NBT.A.2. Fluently add and subtract within 1000 using strategies and algorithms based on place value, properties of operations, and/or the relationship between addition and subtraction.
- 3.OA.A.3. Use multiplication and division within 100 to solve word problems in situations involving equal groups, arrays, and measurement quantities, e.g., by using drawings and equations with a symbol for the unknown number to represent the problem.
- 3.MD.A.1. Tell and write time to the nearest minute and measure time intervals in minutes. Solve word problems involving addition and subtraction of time intervals in minutes, e.g., by representing the problem on a number line diagram.

COMMON CORE STATE STANDARDS FOR ENGLISH LANGUAGE ARTS

READING STANDARDS

- RI.3.1. Ask and answer questions to demonstrate understanding of a text, referring explicitly to the text as the basis for the answers.
- RI.3.3. Describe the relationship between a series of historical events, scientific ideas or concepts, or steps in technical procedures in a text, using language that pertains to time, sequence, and cause/effect.
- RI.3.8. Describe the logical connection between particular sentences and paragraphs in a text (e.g., comparison, cause/effect, first/second/third in a sequence).

WRITING STANDARDS

- W.3.2. Write informative/explanatory texts to examine a topic and convey ideas and information clearly.
- W.3.8. Recall information from experiences or gather information from print and digital sources; take brief notes on sources and sort evidence into provided categories.

SPEAKING AND LISTENING STANDARDS

- SL.3.1. Engage effectively in a range of collaborative discussions (one-on-one, in groups, and teacher-led) with diverse partners on *grade 3 topics and texts*, building on others' ideas and expressing their own clearly.
- SL3.1.D. Explain their ideas and understanding in light of the discussion.

- SL3.3. Ask and answer questions about information from a speaker, offering appropriate elaboration and detail.

FRAMEWORK FOR 21ST CENTURY LEARNING
- Interdisciplinary themes (inventions, history, engineering design process, and progress); Learning and Innovation Skills; Information, Media and Technology Skills; Life and Career Skills

Table 4.6. Key Vocabulary in Lesson 2

Key Vocabulary	Definition
biomimicry	the practice of copying nature to build or improve something
bullet train	high-speed passenger train like those found in Japan
collaboration	working together in groups to achieve a goal or create something
elapsed time	time that passes while an event is occurring
engineering design process	a series of steps that engineers use to create solutions to problems or design and build products
technology	the application of science to improving products or processes.
timeline	a visual representation of events over a period of time in history in which the passage of time is depicted as a straight line
timetable	a chart that shows the arrival and departure times for trains, buses, or airplanes
time zone	a region where a common standard time is used
Transcontinental Railroad	the first train route to stretch across the United States, finished in 1869
westward expansion	the acquisition of territories by the United States, which resulted in the country eventually spanning the entire width of North America, from the Atlantic to the Pacific Ocean

TEACHER BACKGROUND INFORMATION
Trains

The history of rail transportation in the United States is a focus of this lesson, concentrating on passenger transportation. This history is presented in relationship to major historical events in the United States. While students are not expected to have a comprehensive understanding of U.S. history and westward expansion, this lesson provides them with an opportunity to explore landmarks in train travel in the United States and connect this knowledge to the settlement of the nation from coast to coast.

Changes in railway engineering have transformed trains since their first use in the United States in the 1800s. Many of the technological advancements in rail travel are connected with changes in track gauge, or the spacing between rails. In particular, varying track gauges around the country created problems for connecting the growing rail networks in the first half of the 1800s. Although track size was standardized in the late 1800s (4 feet 9 inches, or 1,448 mm), some nonstandard track sizes remain in use to this day for some regional mass transit systems that do not interact with the national railroads. The topic of rail gauge is beyond the scope of this module; however, students should understand that the limiting factor for rail travel in the United States is the availability of railroad tracks.

The following is a brief overview of rail history in the United States (*see www.american-rails.com/railroad-history.html* for more information). Rail history in the United States dates to the early 1800s, when test tracks for steam locomotives began to appear around the country. The birth of the U.S. rail industry is commonly attributed to the inception of the Baltimore & Ohio (B&O) Railroad in 1827. Three years later, the B&O tested a steam-powered locomotive, the Tom Thumb. Rail travel boomed during the 1840s, and by 1850, every state east of the Mississippi except Florida had some railroad track in place. These tracks were largely fragmented and disconnected, however, and rail travel was still considered an experiment and had little government oversight. Between 1850 and 1860, the number of tracks tripled and speeds of trains greatly increased. The trip from Chicago to New York now took two days as opposed to five weeks in 1800 (see *www.mnn.com/green-tech/transportation/stories/how-fast-could-you-travel-across-the-us-in-the-1800s* for more information about the speed of travel in 1800).

Railroads played a key role in the events of the 1860s. The Civil War was highly influenced by the availability of railroad track. Although the country had a rail network of over 30,000 miles in 1861, 70% of these railways were in the northeastern states, leaving the southern states at a disadvantage for moving troops and supplies. The discovery of gold in California prompted proposals to create a railroad that crossed the continental United States, and in 1860, President Abraham Lincoln approved the construction of the Transcontinental Railroad. The Transcontinental Railroad was completed in 1869 and is

considered a pivotal event in the history of the United States, since this was the first time that the nation's two coasts had been connected with a relatively fast and efficient form of transportation. The Transcontinental Railroad was constructed by the Central Pacific and Union Pacific Railroad Companies. These two companies joined their tracks in Utah at an event known as the Golden Spike Ceremony. The railway accelerated the settling of the West, although Native Americans who were angry about the encroachment onto their land often destroyed tracks, making rail travel across the country unreliable and often dangerous.

Before time zones were created, towns across the nation would set their clocks using the noon sun at its highest point as a marker. Local timekeepers would proclaim this time as the official town time, and residents would set their clocks according to this. Railroad travel highlighted the difficulties with this system, since each stop was based on a local time. The railroads responded to the confusion of thousands of unstandardized local times around the country by creating the time zones that are still in use today. To standardize times for railroad travel across the continent, the four major time zones—Eastern, Central, Mountain, and Pacific—were adopted by the railroads in 1883. The divisions were created to avoid populated areas, so they do not always run along straight lines.

By 1900, the United States had over 193,000 miles of rail, and railroads were a primary mode of transportation, with rail depots commonly found in small communities across the country. However, rail travel began to decrease in 1920, as automobile travel became more common and air travel began to establish itself as a major mode of transportation. Although rail travel in general was in decline, the 1930s are remembered for their ornate passenger trains and are known as the streamliner era.

Diesel-electric locomotives became prevalent in the 1930s. In the 1940s, piston-driven steam technology peaked because of bans on building new diesel locomotives due to the resource shortages associated with World War II. After the war, steam locomotives were largely replaced by diesel locomotives, which were more powerful than the diesel-electric locomotives.

Passenger rail travel continued to plummet during the 1950s, and the industry suffered significant losses in its rail passenger service. Many railroads faced financial crises. The plight of railroads was serious enough that in 1967, the federal government stepped in and set up the Consolidated Rail Corporation, composed of bankrupt northeastern rail companies. During this era, the federal government established and partially funded Amtrak, the only nationwide passenger rail service still in use today.

Since 1980, rail travel has begun to reemerge as a critical mode of transportation, particularly in terms of freight transport. Although passenger travel saw declines in recent decades, ridership has begun to rebound, particularly since highways are increasingly congested and high fuel prices force travelers and commuters to look for alternatives. It is notable that, with the exception of the Northeast Corridor

(Boston–Philadelphia–Baltimore–Washington, DC) and some services around Chicago, passenger rail services provided by Amtrak are slower and less available than in most other developed nations around the world.

For more information about the U.S. railroad system and its history, see the following websites:

- *www.american-rails.com/railroad-history.html*

- *www.railswest.com/trainsintro.html*

- *http://amhistory.si.edu/onthemove/themes/story_50_1.html*

- *www.history.com/topics/industrial-revolution/videos/modern-marvels-evolution-of-railroads*

For historical information about international development of rail transportation, see these websites:

- *www.trainhistory.net/railway-history/railroad-history*

- *http://inventors.about.com/od/famousinventions/fl/History-of-the-Railroad.htm*

Engineering and the Engineering Design Process (EDP)

Students begin to gain an understanding of engineering as a profession in this module. In particular, they should understand that engineers are people who design and build products in response to human needs. Engineers apply science and mathematics knowledge to create these designs and solutions. Students should also understand that there are many different types of engineers. For an overview of the various types of engineering professions, see the following websites:

- *www.engineeryourlife.org/?ID=6168*

- *www.nacme.org/types-of-engineering*

- *www.sciencekids.co.nz/sciencefacts/engineering/typesofengineeringjobs.html*

In particular, the following types of engineers design and build trains, train tracks, and train infrastructure and contribute to train technology:

- *Mechanical engineers* design and build mechanical systems (such as motors) and tools.

- *Electrical engineers* design electrical circuits and computer chips.

- *Civil engineers* design bridges, roads, and dams.

- *Aerospace engineers* work mostly on airplanes and space shuttles; however, they also work on high-speed trains, where aerodynamics is especially important.

- *Computer engineers* do work that is similar to that of electrical engineers, but they specialize in computer technology. Much of their work with electrical circuits is on a very small scale, such as in microprocessors.

Students should understand that all these types of engineers need to work in groups to accomplish their work, and that collaboration is important for designing solutions to problems. In this module, students are challenged to work in teams to complete a variety of tasks and to act as train design engineers. They will use the engineering design process (EDP), the same process that professional engineers use in their work. Your students may be familiar with the scientific method but may not have experience with the EDP. Students should understand that the processes are similar but are used in different situations. The scientific method is used to test predictions and explanations about the world. The EDP, on the other hand, is used to create a solution to a problem. In reality, engineers use both processes, and your students' experience will reflect this. They will use the scientific method within the research and knowledge building phase of the EDP as they engage in their inquiry activities and will use the EDP during their final challenge. A good summary of the similarities and differences between the processes can be found at *www.sciencebuddies.org/engineering-design-process/engineering-design-compare-scientific-method.shtml*. An additional resource about the EDP is *www.pbslearningmedia.org/resource/phy03.sci.engin.design.desprocess/what-is-the-design-process*.

A graphic representation of the EDP is attached at the end of this lesson. It may be useful to post this in your classroom. You should be prepared to review each step of the EDP listed on the graphic with students.

Science

Students continue their study of magnets in this lesson by applying their learning from Lesson 1 to a mini design challenge, the Magnificent Magnet Match, in which student teams build and race magnetic boats. Students use the EDP to structure their work during this challenge.

This lesson introduces the concept of biomimicry and its relationship to train engineering. Biomimicry is essentially the practice of emulating nature in designing solutions to human problems and needs. The practice of looking to nature for design inspiration is gaining increasing attention in engineering design, and examples of biomimicry include using termite mounds as models for sustainable architecture, dolphin flippers as models for airplane wings, and mosquito mouths as models for medical needles.

Biomimicry played a role in the design of high-speed trains in Japan. These trains ran through neighborhoods and tunnels, and they were creating sonic booms and noises from their connections to overhead wires. Because these sounds were disturbing people and wildlife, an engineer who was also a birdwatcher looked to the kingfisher, a diving bird that enters the water at high speeds with little splash, for inspiration. The kingfisher's beak provided engineers with an idea for a new shape for the front of a high-speed train that would reduce the sonic boom effect and allow the train to travel faster. For more information, see the following websites:

- *www.greenbiz.com/blog/2012/10/19/how-one-engineers-birdwatching-made-japans-bullet-train-better*

- *http://mi2.org/featured/biomimicry-bullet-trains-innovation*

Mathematics

Students explore the concept of elapsed time in this lesson. They apply their understanding in an exploration of train timetables and the four major time zones across the United States.

For more information about the development of time zones, see the Teacher Background Information section on page 81 and *http://geography.about.com/od/physicalgeography/a/timezones.htm*. For maps of time zones in the United States, see the following websites:

- *http://nationalmap.gov/small_scale/printable/timezones.html#list*

- *www.worldtimezone.com/time-usa12.php*

COMMON MISCONCEPTIONS

Students will have various types of prior knowledge about the concepts introduced in this lesson. Table 4.7 (p. 86) outlines some common misconceptions students may have concerning these concepts. Because of the breadth of students' experiences, it is not possible to anticipate every misconception that students may bring as they approach this lesson. Incorrect or inaccurate prior understanding of concepts can influence student learning in the future, however, so it is important to be alert to misconceptions such as those presented in the table.

Table 4.7. Common Misconceptions About the Concepts in Lesson 2

Topic	Student Misconception	Explanation
Engineering and the engineering design process (EDP)	The term *engineer* always refers to the person who drives a train.	A person who operates a train is known as a train engineer, locomotive engineer, or railroad engineer. These people have training in train operation and are federally licensed as train engineers. A professional who designs and constructs trains is also called an engineer or railroad engineer. These railroad engineers have a specialized combination of training in areas such as electrical, mechanical, and civil engineering.
	Engineers use only the scientific process to solve problems in their work.	The scientific method is used to test predictions and explanations about the world. The EDP, on the other hand, is used to create a solution to a problem. In reality, engineers use both processes. (See Teacher Background Information section on p. 83 for a discussion of this topic.)
Time	A simple subtraction algorithm can be used to calculate elapsed time (for instance, if a movie begins at 1:45 and ends at 3:20, the elapsed time can be calculated as 3:20–1:45).	Students may understand elapsed time best by creating a visual representation such as a timeline or individual student clocks with movable hands. Elapsed time requires counting hours and counting minutes and understanding that there are 60 minutes in 1 hour.
	The time is the same everywhere in the United States.	Local solar time varies by geographic location. As the Earth rotates, different parts of it receive different amounts of sunlight and darkness. Without time zones, noon in one place would be midday in solar terms in some places, early morning in others, and late afternoon in still others. Time zones are based on longitude, with each zone being 15° longitude in width. This means that noon is midday in all locations, but noon in one place will not necessarily correlate with noon in another.

PREPARATION FOR LESSON 2

Review the Teacher Background Information provided, assemble the materials for the lesson, and preview the videos recommended in the Learning Plan Components section below. This lesson involves creating a class timeline. This must be large enough to represent time intervals from the 1800s through 10 years into the future, allowing students to post information on notebook-size pieces of paper along the timeline. It is recommended that this timeline be constructed by attaching the pieces of paper to a wall with masking tape or using clips to hang them on a clothesline strung across the room.

Students design their own scientific inquiry activity in the My Magnetic Question activity (see Introductory Activity/Engagement below and Activity/Exploration, p. 89). After you have reviewed students' questions, you need to prepare appropriate materials for students to carry out their inquiry.

You will introduce students to the EDP during this lesson. You may wish to display the EDP graphic (attached at the end of this lesson) in your classroom for student reference throughout the module. You will assess students on collaboration during this lesson, so you may wish to review the collaboration rubric attached at the end of this lesson.

Students conduct internet research in this lesson. General guidelines for internet research are provided in the Explanation section on page 94. You may also wish to review your school's or district's internet safety policies.

LEARNING PLAN COMPONENTS
Introductory Activity/Engagement

Connection to the Challenge: Begin each day of this lesson by directing students' attention to the driving question for the module and challenge: How can we create a plan and build a prototype for a maglev train to carry passengers to a vacation destination? Hold a brief student discussion of how their learning in the previous days' lessons contributed to their ability to create their plan and build their prototype. You may wish to hold a class discussion, creating a class list of key ideas on chart paper, or have students create a notebook entry with this information.

Social Studies Class: Show a video outlining the history of train technology in the United States (for example, visit YouTube and search for "Trains Through Time" or access the video directly at *www.youtube.com/watch?v=Yh3Ho3vutac*). Hold a class discussion about what changes they saw in the trains pictured in the video. Introduce the idea of a timeline, and create a class timeline showing the progression of the train technologies students noticed in the video, using a large timeline (preferably wall-sized, since you will be adding more information to it). Write student entries on pieces of paper and tape them to the timeline (be sure to leave space to the left of the first entry because you will be adding earlier train history to this timeline; see Elaboration/Application of Knowledge section on p. 95 for details).

Next, ask students to add what they know from personal experience and what they learned in Lesson 1. (Students should understand that there are currently no maglev trains in the United States and only a few high-speed trains.) For example, your timeline might look like the one in Figure 4.1.

Figure 4.1. Sample Timeline of History of Train Travel in the United States

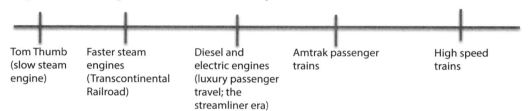

Tom Thumb (slow steam engine)   Faster steam engines (Transcontinental Railroad)   Diesel and electric engines (luxury passenger travel; the streamliner era)   Amtrak passenger trains   High speed trains

Ask students what is missing from the timeline. Guide them to understand that timelines need to have some time frame labeled on them. Have the class guess what years should be assigned to each event, record their guesses, and then have student teams look up each event online to find approximate years for each. (You may want to prompt students to use terms such as "Transcontinental Railroad" and "streamliner era" in their searches.) Tell students that the start of the timeline is 1830, the year Tom Thumb was tested, and the end of the timeline is 10 years in the future. Select students to place the events on the timeline according to this scale.

Science Class: In this lesson, students first create their own inquiry activities related to magnets in the My Magnetic Question activity, and then apply their understanding of magnets and the EDP in a mini design challenge, the Magnificent Magnet Match.

To begin the My Magnetic Question activity, introduce students to the concept of scientific questioning by first offering a statement that starts with "I want to know," such as "I want to know if this ball will bounce more times if I drop it from a higher height." Tell students that scientists start their investigations by wondering about something, but then they change their "I want to know" statements to questions that can be answered. Repeat your "I want to know" statement and ask students how that can be changed into a question. Emphasize that questions should be testable by suggesting a variety of questions, including some that are testable (e.g., "Will the height from which I drop the ball affect the number of times it bounces?") and others that are not testable (e.g., "Is this ball very bouncy?"). Next, ask students which question would be better: "Will the height from which I drop the ball affect the number of times it bounces?" or "How many times will the ball bounce when I drop it from 3 feet?" Point out that while the second question is testable, it doesn't provide much information. Have students work in pairs to devise testable questions about magnets such as the following:

• What solid materials will a magnet work through to pick up a nail?

- How much water will a magnet work through to pick up a nail?

- How does the size of a magnet affect the distance over which it will attract a paper clip?

Review students' questions to ensure that they are testable in the classroom and so that you may prepare appropriate materials.

Mathematics Connection: Tell students that in this lesson they will be learning about time and the way it is used in travel. Show students a sample train timetable from their region (see *www.amtrak.com/train-schedules-timetables*). Demonstrate to students how to read the timetable. Ask students to identify departure and arrival times for various destinations. Use this to segue into a discussion of elapsed time (see Activity/Exploration below).

ELA Connection: Not applicable.

Activity/Exploration

Social Studies Class: Students investigate the role of trains in U.S. history in the context of the Transcontinental Railroad. Tell students that they will look at the role of trains in the growth of the United States in this activity. Ask if they think that we can travel from one side of the country to the other entirely by train. Introduce the idea that this was possible only after the Transcontinental Railroad was built. Students may be familiar with westward expansion from the perspective of pioneers traveling west in covered wagons. Introduce the concept of westward expansion by showing students a U.S. map and telling them that at the beginning of U.S. history, people lived only on the East Coast of the country in the 13 colonies. Ask students to share ideas about why people might have settled here and what might have kept them from spreading out more.

Tell students that in the late 1700s and early 1800s, people wanted more land to farm and hunt, so they began moving west to new, unsettled land. Eventually, people moved all the way across the country to California. When someone in California discovered gold there, many people went to try to find gold; this is called the California gold rush, and over 300,000 people went to California between 1848 and 1855. Ask students how they think all these people got from the East Coast to the West. Guide students to understand that since there were no cars, airplanes, or trains that went across the country, people relied on animal-powered vehicles such as covered wagons. As a class, read and discuss *If You Traveled West in a Covered Wagon*, by Ellen Levine.

Refer to the timeline students created and point out the Transcontinental Railroad. Ask students how this might have changed the experience of moving west. Students should understand that the Transcontinental Railroad made it possible to travel across the country quickly and efficiently for the first time. Show the video "Transcontinental

Railroad" for an overview (visit YouTube and search for "Transcontinental Railroad" or access the video directly at *www.youtube.com/watch?v=P4qYUnm4ZYY*).

Tell students that the Transcontinental Railroad was built by two railroad companies on opposite coasts. The Union Pacific Railroad Company started building tracks in Omaha, Nebraska, and headed west, while the Central Pacific Railroad Company began building in Sacramento, California, and moved east. They eventually met in Promontory Summit, Utah. Have students identify these places on a U.S. map and discuss why it might have been difficult to construct this railroad (point out mountains, rivers, and other land features).

Next, students explore the geography of the route of the Transcontinental Railroad through the Riding the Rails activity.

Riding the Rails

Group students in teams of 2 to 3 for this activity. Students should have devices with internet access and blank U.S. maps (attached at the end of this lesson). Each team of students will be assigned a route that aligns with part of the Transcontinental Railroad. Students will act as "passengers" and should locate the towns between which they are traveling on a map, and then research their route to provide the following information:

- What cities and states did you go through?

- What is the terrain like? Did you cross mountains, rivers, or canyons?

- How did the terrain change on your ride?

- What are some things you saw out the window? (name at least 3)

Assign each of the teams one of the following routes:

- New York, New York, to Chicago, Illinois

- Chicago, Illinois, to Omaha, Nebraska

- Omaha, Nebraska, to Cheyenne, Wyoming

- Cheyenne, Wyoming, to Promontory Summit, Utah

- Promontory Summit, Utah, to Reno, Nevada

- Reno, Nevada, to Alta, California

- Alta, California, to Sacramento, California

Students should complete the Riding the Rails handout (attached at the end of this lesson). After completing the handouts, each team of students should share with the class the following information: where their portion of the trip began and ended, what states they traveled through, and three things they saw out the window.

After students have completed the Riding the Rails activity, remind them of the timeline they made earlier (see the Introductory Activity/Engagement section, p. 87). Point out that the Transcontinental Railroad was built many years ago, and trains at that time were powered by steam (using wood or coal to heat the water). Review with students how trains have changed over the years. Ask students to consider what might be at the end of the timeline, 10 years in the future, for train travel. Have students create a Research Notebook entry with their prediction for the future of passenger train travel. They should include the following information:

- How they predict train travel will change over the next 10 years in their hometown.

- Why they think these changes might happen.

- What they think a train might look like in 10 years (written description and picture).

Science Class: Students plan an investigation to answer the question they devised in the Introductory Activity/Engagement section in the My Magnetic Question activity (see p. 87), and then apply their understanding of magnetism in the Magnificent Magnet Match mini design challenge.

My Magnetic Question

Tell students that they will now answer the questions they posed earlier (see Introductory Activity/Engagement). Tell students that they first need to plan their investigation. Have student pairs work together to create plans for their investigations, using words and pictures, in their STEM Research Notebooks. These plans should be a sequence of steps that students will work through. You may wish to have students use sticky notes, recording one step on each paper, so that they can easily add or remove steps. Facilitate student work by asking guiding questions, encouraging them to think about what kind of data they will collect, how they will collect it, and how they will record data. Be sure that students create data collection charts or tables as well. Review each pair's plans, and when students have a workable plan, instruct them to collect their materials and begin their investigation.

Once students have gathered their data, have student pairs share their findings with the class. To prepare for the presentation, have students create a STEM Research Notebook entry in which they provide the following information:

- Question
- Method of investigation

- What the data tells about the question (What are the results?)
- Conclusions

Magnificent Magnet Match

Ask students to brainstorm about who they think designs and builds trains and train tracks. Record their answers. Introduce the idea that engineers are people who design and build things to solve problems or fill human needs. Students should understand that there are many different kinds of engineers (see Teacher Background Information section on p. 83 for an overview) and that these engineers do different kinds of work than the people who drive the trains, who are often called train engineers.

Remind students that they are going to act as engineers in this module as they build their Maglevacation Trains. To act as engineers, they need to know how engineers work. Introduce the idea that engineers work together in teams to solve problems and use a process called the engineering design process, or EDP, to do their work.

Show the video "The Engineering Process: Crash Course Kids #12.2," which explains and gives examples of the steps of the EDP (visit YouTube and search for "The Engineering Process: Crash Course Kids #12.2" or access directly at *www.youtube.com/ watch?v=fxJWin195kU*). After the video, show students the EDP graphic (attached at the end of this lesson) and tell them that they will use this process as they work in the teams they formed in Lesson 1 to solve their Maglevacation Train Challenge. Tell students that they will now have the chance to practice using the steps of the EDP and to put into action what they've learned about magnets. They will apply their learning about magnets and experience the EDP as they participate in a mini design challenge called the Magnificent Magnet Match.

In this activity, students will be challenged to apply their understanding of magnetic poles and fields to create boats "powered" by magnetism. Each team will be provided with a set of materials, a list of rules, and EDP handouts. The goal is to create a boat that can move across a body of water (wading pool) as fast as possible without being touched by human hands. The solution to their challenge is fairly simple; however, they must use the EDP and meet the requirements of the challenge. *Safety note:* Be sure to immediately wipe up any splashed or spilled water, which could cause someone to slip and fall!

Tell students that they are being challenged to create a vehicle to participate in a race to cross a body of water the fastest. Read them the rules for the challenge:

- You can use any of the materials provided but do not need to use them all.
- Your vehicle must cross the water as quickly as possible.
- You may not touch the vehicle with your hands except to put it into the water at the start of the race.

- Your vehicle must be unique and distinguishable from other teams' vehicles, and you may decorate it however you like.

- You must use the EDP to design your vehicle and fill out the Engineer It! handouts

- Everyone on the team must participate in creating the vehicle.

- Each team will be allowed one test run in the wading pool before the class race.

- You will have ____ minutes to design and test your vehicle. (Assign an amount of time for teams to work on their vehicles; remember to allow enough time for a class race).

Each student should use the Engineer It! handouts to track the team's progress. Students will probably use a variation of a design in which a magnet is attached to the soap with rubber bands, using another magnet to push or pull the boat.

After students have completed the design, build, and test phases, hold a class race. Next, have a class discussion about how some of the boats varied in design. Ask students questions to help guide the discussion: What worked well and what didn't? How did your team work together in this activity?

Mathematics Connection: In the Introductory Activity/Engagement phase, students explored train timetables and how to read these. Ask students how they know how much time it will take to get somewhere. Introduce the concept of elapsed time as the amount of time between the start and finish of an event—in other words, how long something takes. Introduce calculations for elapsed time. Be sure that students can work easily with hours and minutes. The following video provides an overview of these calculations: *www.youtube.com/watch?v=zXFZUMjehDU*.

Refer back to the arrival and departure times students identified for various destinations in the Introductory Activity (see p. 89), and have students calculate elapsed times. Students should include these calculations in their STEM Research Notebooks. You may also wish to distribute clock templates on which students can draw times (for free printable clock templates see *www.template.net/design-templates/print/printable-clock-template*). An interactive activity for elapsed time is available at *www.splashmath.com/math-skills/third-grade/time/elapsed-time*.

ELA Connection: Introduce the notion of biomimicry to students, and have them brainstorm ideas in their Research Notebooks about why engineers might look to nature for design ideas.

Have students read *From Kingfishers to … Bullet Trains,* by Wil Mara. Students should look for the various steps of the EDP while reading and create a STEM Research Notebook

entry in which they note how each of these steps was used in the process of creating the Japanese bullet train.

Explanation

This section contains an overview of concepts students should understand in this lesson. You may also wish to have students share what they have learned and explain their understanding of lesson concepts.

Social Studies Class: Explain to students that a timeline is graphic representation of time expressed as a straight line. Timelines are often used to represent historical events in chronological order, and the spaces between events on a timeline represent amounts of elapsed time.

Tell students that there were no railroads in the United States from the time of the earliest settlers up until the early 1800s. Before the advent of railroads in the United States, all land travel was by foot, horseback, or horse-drawn vehicle. You may wish to link this information to the geography of the original 13 colonies, which were relatively small in size and clumped together on the East Coast of the country. Point out how much larger midwestern and western states are in comparison, and lead students to understand that rail travel allowed people to travel large distances at an unprecedented rate of speed.

Provide students with some basic guidelines on conducting internet searches for information. First, direct them to use a safe and reliable search engine (if your school uses a particular kid-friendly search engine, direct students there). Then, encourage them to use the following steps:

1. Enter one or more key words in the search bar, and search for these terms.

2. Skim the list of sites that appears.

3. Preview sites that you think may be helpful by going to the site and quickly reviewing the information that is there.

4. Spend your time on sites that your preview shows may be helpful.

5. Focus on credible sites (e.g., government websites or educational websites from known sources) instead of blogs and sources that you don't recognize.

6. Try using different search terms if your first search didn't lead you to the information you want.

Science Class: Review the steps of the EDP with students before they begin the mini design challenge, Magnificent Magnet Match. You may also wish to review the basics of magnetism with students, including how poles of magnets interact (opposites attract) and how magnets can be used for movement.

Mathematics Connection: Explain to students that train timetables are published as charts and that they will read them like they read other charts. There is a great deal of information on these timetables, so you may need to help them interpret the departure and arrival times while reading the charts across and down.

ELA Connection: Students will have the opportunity in the next section (Elaboration/Application of Knowledge) to apply their understanding of biomimicry by creating an advertising brochure for a fictional product or item that is modeled on a feature in nature. In this Explanation phase, review the procedures for planning writing. Consider providing students with a graphic organizer that allows them to plan their main ideas and supporting details (for printable graphic organizers for classroom use see *www.eduplace.com/ graphicorganizer*).

Elaboration/Application of Knowledge

Social Studies Class: Students can apply their understanding of the history of rail travel, the history of the United States, and geography to investigate the prevalence of passenger rail travel across the country. Assign each team of 2 to 3 students a region of the country, and have them research how available rail travel is for both cross-country and regional travel.

Since Amtrak is the primary passenger rail carrier in the nation, students can access this information on the Amtrak website. A map of the Amtrak system is available at *www.amtrak.com/ccurl/948/674/System0211_101web,0.pdf*, and an interactive route map is available at *www.amtrak.com/find-train-bus-stations-train-routes.* A state-by-state summary of passenger rail travel for a number of states is available at *www.american-rails.com/ passenger-train-travel.html.*

Extend students' understanding of train history by adding to the timeline the class created earlier (see the Introductory Activity/Engagement section, p. 87). Ask students to share their ideas about the invention of trains. Do they think that trains were invented in the United States? Tell students that the first working steam locomotive was actually built in 1804 by an Englishman named Richard Trevithick. Add this to the timeline. Then, introduce the idea of tramways to students. Wagon tramways were the precursor of modern railways. These vehicles were pulled by people or animals and had wheels that ran in grooves in stone or on wooden planks to help the vehicles stay on course. These tramways were used in Germany in the 1500s (around 1550–1560). Add these items to the train history timeline. Ask students to reflect on whether the timeline's title is still correct. Ask them to propose another title for the timeline (for example, History of Trains). Continue to add to this timeline throughout the module as students learn about trains and train technology.

Science Class: Have students create a STEM Research Notebook entry in response to the prompt provided next.

STEM Research Notebook Prompt

Students should respond to the following prompt in their Research Notebooks: *In this lesson, you learned about working the way professional engineers do, using the engineering design process. Describe your team's experience using this process. What went well? What did not go well? How do you think your team worked together?*

Mathematics Connection: Students who have an understanding of elapsed time and time calculations may extend this understanding to the four time zones across the United States. Explain to students the connection between time zones and railroads, and show them the map of U.S. time zones (attached at the end of this lesson). Show the following video to give an overview of the history and use of time zones: *www.history.com/topics/industrial-revolution/videos/setting-time-zones.*

Have students apply their understanding of elapsed time by calculating the time difference between their time zone and the three other time zones in the country, using the U.S. time zone map. Using this information, assign students a variety of "home times," and ask them to calculate the time in various parts of the country. You may wish to incorporate map skills in this lesson by naming major cities, states, or landmarks (for instance, the Grand Canyon or Lake Erie) and allowing students to locate these on the map.

ELA Connection: Based on their understanding of biomimicry and engineering design from the book *From Kingfishers to … Bullet Trains,* have students conceive a fictional product based on a feature from nature. Students should plan and create an advertising brochure for this product, highlighting its usefulness, the benefits of its design, and how it is connected to a feature in nature. Students can then present their product ideas to the class, using the information included in their brochures.

Evaluation/Assessment

Students may be assessed on the following performance tasks and other measures listed.

Performance Tasks

- Riding the Rails handouts
- My Magnetic Question research question
- My Magnetic Question presentation
- Magnificent Magnet Match EDP handouts
- Elapsed time calculations Research Notebook entry

Other Measures

- Other STEM Research Notebook entries

- Collaboration (see Collaboration Rubric attached at the end of this lesson)
- Participation in class and group discussions

INTERNET RESOURCES

Overview of rail history in the United States

- *www.american-rails.com/railroad-history.html*
- *www.railswest.com/trainsintro.html*
- *http://amhistory.si.edu/onthemove/themes/story_50_1.html*
- *www.history.com/topics/industrial-revolution/videos/modern-marvels-evolution-of-railroads*

Overview of the Transcontinental Railroad

- *www.youtube.com/watch?v=P4qYUnm4ZYY*

History of international development of rail transportation

- *www.trainhistory.net/railway-history/railroad-history*
- *http://inventors.about.com/od/famousinventions/fl/History-of-the-Railroad.htm*

Speed of travel in 1800, Mother Nature Network

- *www.mnn.com/green-tech/transportation/stories/how-fast-could-you-travel-across-the-us-in-the-1800s*

Video showing train technology over time

- *www.youtube.com/watch?v=Yh3Ho3vutac*

Overview of the various types of engineering professions

- *www.engineeryourlife.org/?ID=6168*
- *www.nacme.org/types-of-engineering*
- *www.sciencekids.co.nz/sciencefacts/engineering/typesofengineeringjobs.html*

Summary of differences between EDP and scientific process

- *www.sciencebuddies.org/engineering-design-process/engineering-design-compare-scientific-method.shtml*

Additional resources about the EDP

- *www.pbslearningmedia.org/resource/phy03.sci.engin.design.desprocess/what-is-the-design-process*
- *www.youtube.com/watch?v=fxJWin195kU*

Information about biomimicry

- *https://biomimicry.org/biomimicry-examples*
- *www.greenbiz.com/blog/2012/10/19/how-one-engineers-birdwatching-made-japans-bullet-train-better*
- *http://mi2.org/featured/biomimicry-bullet-trains-innovation*

Information about the development and use of time zones

- *http://geography.about.com/od/physicalgeography/a/timezones.htm*
- *www.history.com/topics/industrial-revolution/videos/setting-time-zones*

Maps of time zones in the United States

- *http://nationalmap.gov/small_scale/printable/timezones.html#list*
- *www.worldtimezone.com/time-usa12.php*

Elapsed time calculations

- *www.youtube.com/watch?v=zXFZUMjehDU*
- *www.splashmath.com/math-skills/third-grade/time/elapsed-time*

Amtrak system map and interactive route map

- *www.amtrak.com/ccurl/948/674/System0211_101web,0.pdf*
- *www.amtrak.com/find-train-bus-stations-train-routes*

State-by-state summary of passenger rail travel

- *www.american-rails.com/passenger-train-travel.html*

Printable clock templates

- *www.template.net/design-templates/print/printable-clock-template*

Printable graphic organizers

- *www.eduplace.com/graphicorganizer*

Names of Group Members: _____

STUDENT HANDOUT, PAGE 1

RIDING THE RAILS

Our railroad route was from _____ to _____ .

We traveled through the following states: _____

Some cities we traveled through were _____

We traveled about this many miles: _____

The terrain was (describe what the landscape was like—for example, were there mountains, rivers, canyons?) _____

We saw this out the window (draw and label a picture, using the back of this page if you need to): _____

Transportation in the Future Lesson Plans

STUDENT HANDOUT, PAGE 2

RIDING THE RAILS

Name:

Mark your route on this map, and label the cities and states you passed through.

NATIONAL SCIENCE TEACHERS ASSOCIATION

Name: _____ Team Name: _____

STUDENT HANDOUT, PAGE 1

MAGNIFICENT MAGNET MATCH

Step 1: (Define) State the problem (what are you trying to do?).

Step 2: (Learn) What solutions can you and your team imagine?

Step 3: (Plan) Sketch your design here and label materials. Be sure to consider safety issues when planning your design.

Step 4: (Try) Build it! Did everyone on your team have a chance to help?

Transportation in the Future Lesson Plans

Name: _____   Team Name: _____

STUDENT HANDOUT, PAGE 2

MAGNIFICENT MAGNET MATCH

Step 5: (Test)   Test your design. How did it work? Could it work better?

Step 6: (Decide)   This is your chance to use your test results to decide what changes to make. What did you change?

Step 7: Share your design! Who will race your vehicle in the class competition?

U.S. STANDARD TIME ZONES

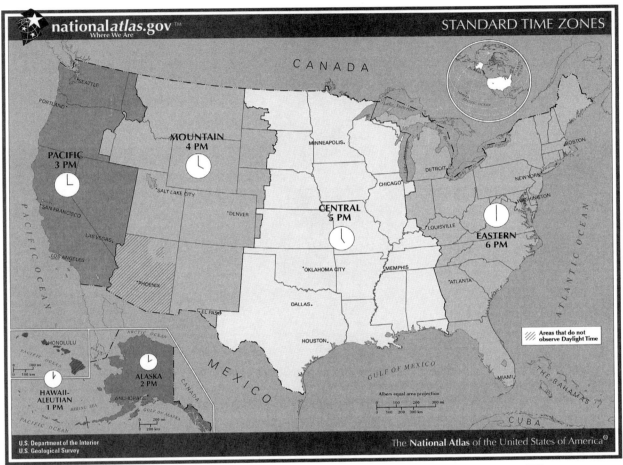

Source: U.S. Department of the Interior, *http://nationalmap.gov/small_scale/printable/timezones.html#list.*

ENGINEERING DESIGN PROCESS

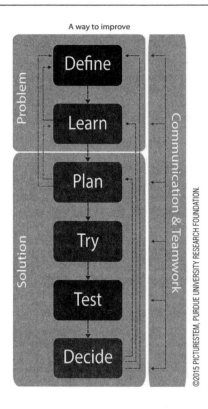

Engineer It! Magnificent Magnet Match Collaboration Rubric (30 points possible)

Student Name: _____   Team Name: _____

Individual Performance	Below Standard (0–3)	Approaching Standard (4–7)	Meets or Exceeds Standard (8–10)	Student Score
INDIVIDUAL ACCOUNTABILITY	• Student is unprepared. • Student does not communicate with team members and does not manage tasks as agreed on by the team. • Student does not complete or participate in project tasks. • Student does not complete tasks on time. • Student does not use feedback from others to improve work.	• Student is usually prepared. • Student sometimes communicates with team members and manages tasks as agreed on by the team, but not consistently. • Student completes or participates in some project tasks but needs to be reminded. • Student completes most tasks on time. • Student sometimes uses feedback from others to improve work.	• Student is consistently prepared. • Student consistently communicates with team members and manage tasks as agreed on by the team. Student discusses and reflects on ideas with the team. • Student completes or participates in project tasks without being reminded. • Student completes tasks on time. • Student uses feedback from others to improve work.	
TEAM PARTICIPATION	• Student does not help the team solve problems; may interfere with teamwork. • Student does not express ideas clearly, pose relevant questions, or participate in group discussions. • Student does not give useful feedback to other team members. • Student does not volunteer to help others when needed.	• Student cooperates with the team but may not actively help solve problems. • Student sometimes expresses ideas, poses relevant questions, elaborates in response to questions, and participates in group discussions. • Student provides some feedback to team members. • Student sometimes volunteers to help others.	• Student helps the team solve problems and manage conflicts. • Student makes discussions effective by clearly expressing ideas, posing questions, and responding thoughtfully to team members' questions and perspectives. • Student gives useful feedback to others so they can improve their work. • Student volunteers to help others if needed.	

(continued)

Individual Performance	Below Standard (0–3)	Approaching Standard (4–7)	Meets or Exceeds Standard (8–10)	Student Score
PROFESSIONALISM AND RESPECT FOR TEAM MEMBERS	• Student is impolite or disrespectful to other team members. • Student does not acknowledge or respect others' ideas and perspectives.	• Student is usually polite and respectful to other team members. • Student usually acknowledges and respects others' ideas and perspectives.	• Student is consistently polite and respectful to other team members. • Student consistently acknowledges and respects others' ideas and perspectives.	

TOTAL SCORE: _____

Lesson Plan 3: Portable People

In this lesson, students explore passenger train travel today and compare it with other modes of transportation. Using the jigsaw learning method, students convene in expert groups to gather data on various aspects of train travel, automobile travel, and air travel within the United States. Design teams then act as "travel agents," using the findings from the expert groups regarding the disadvantages and benefits of each mode of travel to decide on the best mode of transportation to the Maglevacation Train Challenge destination each team chose in Lesson 1 and making recommendations to fictional clients. In addition, students conduct research into the geography, climate, and culture of their destinations to create electronic travel brochures. Each team creates a presentation using technology such as PowerPoint to share its findings about its chosen mode of transportation and vacation destination with the class. In science, students explore electromagnets and identify independent and dependent variables and formulate hypotheses. Students apply their understanding of the EDP to create a balloon-propelled vehicle and hypothesize about and test the relationship between weight and acceleration for their vehicles. Mathematics connections include creating a budget for the fictional clients based on the given destination. Students delve more deeply into the technology used in maglev trains and explore the advantages of maglev trains for the future of transportation.

ESSENTIAL QUESTIONS

- What are the advantages of various modes of passenger transportation?
- What are the disadvantages of various modes of transportation?
- How do we use maps to predict factors such as climate?
- How do we create a budget?
- How do magnets contribute to the movement of maglev trains?
- How do we identify independent and dependent variables?
- How does weight influence motion?

ESTABLISHED GOALS AND OBJECTIVES

At the conclusion of this lesson, students will be able to do the following:

- Use their understanding of the advantages and disadvantages of various modes of transportation to make recommendations about the preferred method of travel to a particular destination

- Use their understanding of geography and maps to predict climate conditions in various locations around the United States

- Apply their understanding of budgets to create a travel budget

- Create a persuasive argument for visiting their chosen vacation destination

- Create a compelling presentation using technology highlighting their findings

- Apply findings from research to calculate costs per mile

- Demonstrate a conceptual understanding of electromagnets

- Identify independent and dependent variables in an investigation

- Formulate hypotheses and test these hypotheses

- Understand that objects must have a force applied to them to initiate movement and demonstrate this understanding by creating a vehicle propelled by air leaving a balloon

- Understand and demonstrate that the weight of an object affects how much force is needed to initiate motion

- Discuss how magnets are used in maglev trains and predict how they could create their own prototype maglev vehicle

TIME REQUIRED

- 7 days (approximately 45 minutes each day; see Tables 3.7–3.8)

MATERIALS

Required Materials for Lesson 3

- STEM Research Notebooks

- Access to internet technology for showing video clips and for student research

- Classroom map of the United States

- Chart paper

- Handouts (attached at the end of this lesson)

Additional Materials for Mathematics Budgeting

- 15–20 boxes or bags

- "Receipts" for student purchases (small slips of paper with items and prices)
- Play money $10 bills (20 per student)

Additional Materials for Let's Go!

- Access to technology such as PowerPoint for student presentations, if possible
- Let's Go! handouts (1 per student)

Additional Materials for Electromagnetic Demonstration (per class unless otherwise indicated):

- Safety glasses or goggles (1 pair per student)
- 1 lantern battery
- 3 lengths of copper wire: 24, 36, and 48 inches
- 3 nails, 6–9 inches long
- 10–15 paper clips

Additional Materials for Balloon Car (per group unless otherwise indicated):

- Safety glasses or goggles (1 pair per student)
- 2 foam trays (grocery store meat trays about 6 × 8 inches work well) or 2 pieces of corrugated cardboard of the same size
- Balloon Car template (attached at the end of this lesson)
- Scissors
- Masking tape
- Flexible drinking straw
- Balloon
- 6 straight pins or toothpicks (straight pins will work better with Styrofoam)
- 10 pennies or small weights, such as washers
- Meter stick
- Timer or stopwatch
- Balloon Car handout (1 per student)

CONTENT STANDARDS AND KEY VOCABULARY

Table 4.8 lists the content standards from the *NGSS, CCSS,* and Framework for 21st Century Learning that this lesson addresses, and Table 4.9 (p. 115) presents the key vocabulary. Vocabulary terms are provided for both teacher and student use. Teachers may choose to introduce some or all of the terms to students.

Table 4.8. Content Standards Addressed in STEM Road Map Module Lesson 3

NEXT GENERATION SCIENCE STANDARDS

PERFORMANCE EXPECTATIONS

- 3-PS2-4. Define a simple design problem that can be solved by applying scientific ideas about magnets.

- 3-5-ETS1-2. Generate and compare multiple possible solutions to a problem based on how well each is likely to meet the criteria and constraints of the problem.

- 3-5-ETS1-3. Plan and carry out fair tests in which variables are controlled and failure points are considered to identify aspects of a model or prototype that can be improved.

SCIENCE AND ENGINEERING PRACTICES

Asking Questions and Defining Problems

- Ask questions about what would happen if a variable is changed.

- Identify scientific (testable) and non-scientific (non-testable) questions.

- Ask questions that can be investigated and predict reasonable outcomes based on patterns such as cause and effect relationships.

- Use prior knowledge to describe problems that can be solved.

- Define a simple design problem that can be solved through the development of an object, tool, process, or system and includes several criteria for success and constraints on materials, time, or cost.

Obtaining, Evaluating, and Communicating Information

- Read and comprehend grade-appropriate complex texts and/or other reliable media to summarize and obtain scientific and technical ideas and describe how they are supported by evidence.

- Compare and/or combine across complex texts and/or other reliable media to support the engagement in other scientific and/or engineering practices.

- Combine information in written text with that contained in corresponding tables, diagrams, and/or charts to support the engagement in other scientific and/or engineering practices.

- Obtain and combine information from books and/or other reliable media to explain phenomena or solutions to a design problem.

- Communicate scientific and/or technical information orally and/or in written formats, including various forms of media as well as tables, diagrams, and charts.

Developing and Using Models

- Identify limitations of models.

- Collaboratively develop and/or revise a model based on evidence that shows the relationships among variables for frequent and regular occurring events.

- Develop a model using an analogy, example, or abstract representation to describe a scientific principle or design solution.

- Develop and/or use models to describe and/or predict phenomena.

- Develop a diagram or simple physical prototype to convey a proposed object, tool, or process.

- Use a model to test cause and effect relationships or interactions concerning the functioning of a natural or designed system.

Planning and Carrying Out Investigations

- Plan and conduct an investigation collaboratively to produce data to serve as the basis for evidence, using fair tests in which variables are controlled and the number of trials considered.

- Evaluate appropriate methods and/or tools for collecting data.

- Make observations and/or measurements to produce data to serve as the basis for evidence for an explanation of a phenomenon or test a design solution.

- Make predictions about what would happen if a variable changes.

- Test two different models of the same proposed object, tool, or process to determine which better meets criteria for success.

Constructing Explanations and Designing Solutions

- Use evidence (e.g., measurements, observations, patterns) to construct or support an explanation or design a solution to a problem.

- Identify the evidence that supports particular points in an explanation.

- Apply scientific ideas to solve design problems.

- Generate and compare multiple solutions to a problem based on how well they meet the criteria and constraints of the design solution.

Analyzing and Interpreting Data

- Analyze data to refine a problem statement or the design of a proposed object, tool, or process.

- Use data to evaluate and refine design solutions.

Using Mathematical and Computational Thinking

- Decide if qualitative or quantitative data are best to determine whether a proposed object or tool meets criteria for success.

- Describe, measure, estimate, and/or graph quantities (e.g., area, volume, weight, time) to address scientific and engineering questions and problems.

Constructing Explanations and Designing Solutions

- Use evidence (e.g., measurements, observations, patterns) to construct or support an explanation or design a solution to a problem.
- Identify the evidence that supports particular points in an explanation.
- Apply scientific ideas to solve design problems.
- Generate and compare multiple solutions to a problem based on how well they meet the criteria and constraints of the design solution.

DISCIPLINARY CORE IDEAS

PS2.A: Forces and Motion

- Each force acts on one particular object and has both strength and a direction. An object at rest typically has multiple forces acting on it, but they add to give zero net force on the object. Forces that do not sum to zero can cause changes in the object's speed or direction of motion.
- The patterns of an object's motion in various situations can be observed and measured; when that past motion exhibits a regular pattern, future motion can be predicted from it.

PS2.B: Types of Interactions

- Objects in contact exert forces on each other.

ETS1.A: Defining and Delimiting Engineering Problems

- Possible solutions to a problem are limited by available materials and resources (constraints). The success of a designed solution is determined by considering the desired features of a solution (criteria). Different proposals for solutions can be compared on the basis of how well each one meets the specified criteria for success or how well each takes the constraints into account. (3-5-ETS1-1)

ETS1.B: Developing Possible Solutions

- Research on a problem should be carried out before beginning to design a solution. Testing a solution involves investigating how well it performs under a range of likely conditions. (3-5-ETS1-2)
- At whatever stage, communicating with peers about proposed solutions is an important part of the design process, and shared ideas can lead to improved designs. (3-5-ETS1-2)
- Tests are often designed to identify failure points or difficulties, which suggest the elements of the design that need to be improved. (3-5-ETS1-3)

ETS1.C: Optimizing the Design Solution

- Different solutions need to be tested in order to determine which of them best solves the problem, given the criteria and the constraints. (3-5-ETS1-3)

CROSSCUTTING CONCEPTS

Patterns
- Similarities and differences in patterns can be used to sort, classify, communicate, and analyze simple rates of change for natural phenomena and designed products.

Scale, Proportion, and Quantity
- Standard units are used to measure and describe physical quantities such as weight, time, temperature, and volume.

Energy and Matter
- Energy can be transferred in various ways and between objects.

Cause and Effect
- Cause and effect relationships are routinely identified, tested, and used to explain change.

Stability and Change
- Change is measured in terms of differences over time and may occur at different rates.

Influence of Science, Engineering, and Technology on Society and the Natural World
- Engineers improve existing technologies or develop new ones to increase their benefits, decrease known risks, and meet societal demands.

COMMON CORE STATE STANDARDS FOR MATHEMATICS

MATHEMATICAL PRACTICES
- MP1. Make sense of problems and persevere in solving them.
- MP2. Reason abstractly and quantitatively.
- MP4. Model with mathematics.
- MP5. Use appropriate tools strategically.
- MP6. Attend to precision.

MATHEMATICAL CONTENT
- 3.NBT.A.2. Fluently add and subtract within 1000 using strategies and algorithms based on place value, properties of operations, and/or the relationship between addition and subtraction.

- 3.OA.A.3. Use multiplication and division within 100 to solve word problems in situations involving equal groups, arrays, and measurement quantities, e.g., by using drawings and equations with a symbol for the unknown number to represent the problem.

- 3.MD.A.1. Tell and write time to the nearest minute and measure time intervals in minutes. Solve word problems involving addition and subtraction of time intervals in minutes, e.g., by representing the problem on a number line diagram.

COMMON CORE STATE STANDARDS FOR ENGLISH LANGUAGE ARTS

READING STANDARDS

- RI.3.1. Ask and answer questions to demonstrate understanding of a text, referring explicitly to the text as the basis for the answers.

- RI.3.3. Describe the relationship between a series of historical events, scientific ideas or concepts, or steps in technical procedures in a text, using language that pertains to time, sequence, and cause/effect.

- RI.3.8. Describe the logical connection between particular sentences and paragraphs in a text (e.g., comparison, cause/effect, first/second/third in a sequence).

WRITING STANDARDS

- W.3.1. Write opinion pieces on topics or texts, supporting a point of view with reasons.

- W.3.1.A. Introduce the topic or text they are writing about, state an opinion, and create an organizational structure that lists reasons.

- W.3.1.B. Provide reasons that support the opinion.

- W.3.1.C. Use linking words and phrases (e.g., *because, therefore, since, for example*) to connect opinion and reasons.

- W.3.7. Conduct short research projects that build knowledge about a topic.

- W.3.8. Recall information from experiences or gather information from print and digital sources; take brief notes on sources and sort evidence into provided categories.

SPEAKING AND LISTENING STANDARDS

- SL.3.1. Engage effectively in a range of collaborative discussions (one-on-one, in groups, and teacher-led) with diverse partners on *grade 3 topics and texts,* building on others' ideas and expressing their own clearly.

- SL3.1.D. Explain their ideas and understanding in light of the discussion.

- SL3.3. Ask and answer questions about information from a speaker, offering appropriate elaboration and detail.

- SL3.4. Report on a topic or text, tell a story, or recount an experience with appropriate facts and relevant, descriptive details, speaking clearly at an understandable pace.

- SL.3.6. Speak in complete sentences when appropriate to task and situation in order to provide requested detail or clarification.

FRAMEWORK FOR 21ST CENTURY LEARNING

Interdisciplinary themes (inventions, history, engineering design process, and progress); Learning and Innovation Skills; Information, Media and Technology Skills; Life and Career Skills

Table 4.9. Key Vocabulary in Lesson 3

Key Vocabulary	Definition
atmospheric science	the study of the earth's atmosphere, climate, and weather and how human activity and natural processes affect these
budget	an estimate of expected expenses and available money over a period of time
climate	the usual weather conditions in an area
cost analysis	a summary of the costs related to an activity
data	facts collected in order to understand something
environmental factors	any actions or conditions that affect living things or communities
modes of transportation	different ways of transporting something or someone
persuasive language	the use of language to influence people to change their minds about something or agree with one's opinion
propulsion	the act of moving something forward
travel agent	a person who makes travel arrangements for clients

TEACHER BACKGROUND INFORMATION

In this lesson, students compare various modes of transportation across several factors using the jigsaw learning method (see below). Students also complete the first component of the Maglevacation Train Challenge. This task requires students to act as travel agents, with each team collecting data about its chosen vacation destination and using this information to create a persuasive argument for fictional clients to visit this destination.

Jigsaw Learning Method

Students are also tasked with comparing various modes of transportation across several factors. To do this, you should employ the jigsaw method of grouping students. The jigsaw method involves dividing students into small groups, called expert groups in this lesson. Each expert group is assigned one factor of a larger phenomenon to investigate. Expert groups then present their findings to the larger group, so that together, the class forms an overview of the various factors, using the presentations to put them together, much the same way as a puzzle is constructed. For more information on the jigsaw method, see *www.jigsaw.org/overview*.

For this lesson, each expert group is charged with investigating one factor of travel across three modes of transportation: airplane, train, and automobile travel. You should ensure that each design team has a member on each of the expert teams, so that each design team has an "expert" on each factor represented; thus, the number of expert groups should match the number of members on each design team as closely as possible. Suggested factors for expert groups to investigate across the three modes of transportation include cost, availability and convenience, environmental factors, passenger comfort, and safety.

Career Connections

As career connections related to this lesson, you may wish to introduce the following:

- *Travel Agent:* Travel agents arrange transportation and lodging for clients. They may offer advice on various travel destinations, create itineraries, make reservations, and provide information or access to recreational activities. For more information, see *www.bls.gov/ooh/sales/travel-agents.htm#tab-2.*

- *Geographer:* Geographers study the Earth's natural land formations and human society, with a focus on the relationship between these phenomena. In particular, they study the characteristics of various parts of the Earth, including physical characteristics and human culture. Many geographers work for the federal government. Teaching and field research are other areas in which geographers work. For more information, see *www.bls.gov/ooh/life-physical-and-social-science/ geographers.htm.*

- *Atmospheric Scientist:* Atmospheric scientists study the weather and climate. This field encompasses several types of careers, including climatologists, meteorologists, and broadcast meteorologists (see specifics on each below). For more information, see *www.bls.gov/ooh/life-physical-and-social-science/atmospheric-scientists-including-meteorologists.htm#tab-2.*

- *Climatologist:* Climatologists study historical weather patterns to identify long-term weather patterns or changes in climate factors and use these data to make predictions, such as anticipated precipitation levels.

- *Meteorologist:* Meteorologists have training in weather patterns and forecasting techniques. They produce weather reports that can be used by the general public or by specific groups, such as farmers or airports.

- *Broadcast meteorologist:* Broadcast meteorologists provide information to the public on current and expected weather conditions for specific locations. They may or may not have training as meteorologists.

Maglev Trains

This lesson will feature more in-depth discussions about maglev trains and their future in the United States. For an overview of maglev train technology, see *http://science. howstuffworks.com/transport/engines-equipment/maglev-train.htm*. For a discussion of why maglev trains are not currently used in the United States, see *www.cnn.com/2015/05/03/ opinions/smart-high-speed-trains-america*.

COMMON MISCONCEPTIONS

Students will have various types of prior knowledge about the concepts introduced in this lesson. Table 4.10 outlines some common misconceptions students may have concerning these concepts. Because of the breadth of students' experiences, it is not possible to anticipate every misconception that students may bring as they approach this lesson. Incorrect or inaccurate prior understanding of concepts can influence student learning in the future, however, so it is important to be alert to misconceptions such as those presented in the table.

Table 4.10. Common Misconceptions About the Concepts in Lesson 3

Topic	Student Misconception	Explanation
Forces	Nonmoving objects do not exert a force.	Stationary objects can exert forces on other objects. For example, when you roll a toy car over a tabletop, there is a frictional force between the nonmoving tabletop and the car wheels.
	If a moving object is slowing, the force that was propelling it forward is decreasing.	When a force acts on a moving object in a direction opposite the direction of motion, the moving object will slow, even if the force that was propelling the object forward continues.

PREPARATION FOR LESSON 3

Review the Teacher Background Information provided, assemble the materials for the lesson, and preview the videos recommended in the Learning Plan Components section below. The Mathematics Budgeting activity requires some preparation and setup before students arrive; see this activity on page 119 in the Introductory Activity/Engagement section for more details. Assessment of the Let's Go! activity will use both the collaboration rubric (one per student) and the presentation rubric (one per group). You may wish to review these rubrics (attached at the end of this lesson).

The activities included in the Activity/Exploration section of this lesson (see p. 121) require students to conduct research in teams. Using the EDP will help students organize and structure their work. You may wish to schedule a class session with your school librarian or media specialist about conducting research. You may also wish to consider inviting parent volunteers or older students to work with individual teams as they undertake these activities. Alternatively, you may choose to prepare summaries of information to provide to students from the websites suggested, since the reading level of some sites may be significantly above a third-grade level (see the Activity/Exploration and Explanation sections on pp. 121 and 127).

LEARNING PLAN COMPONENTS
Introductory Activity/Engagement

Connection to the Challenge: Begin each day of this lesson by directing students' attention to the driving question for the module and challenge: How can we create a plan and build a prototype for a maglev train to carry passengers to a vacation destination? Hold a brief student discussion of how their learning in the previous days' lessons contributed to their ability to create their plan and build their prototype. You may wish to hold a class discussion, creating a class list of key ideas on chart paper, or you may wish to have students create a notebook entry with this information.

Social Studies Class: Tell students that while they have concentrated on train travel up until now, in this lesson they are going to consider how train travel compares with other types of transportation. Have students brainstorm a list of types of transportation. You may wish to use a KWL chart for this purpose.

Show the following video, which gives an overview of forms of transportation throughout history: *www.geekwire.com/2015/planes-trains-and-automobiles-this-video-is-a-fun-brief-look-at-the-history-of-transportation*. Ask students to name some forms of transportation in the video that were new to them and complete the KWL chart. Point out to them that maglev trains were in the video and that they have actually been around since 1984. This is a good time to remind students that, although maglev trains are used in

Europe, Japan, and China, there are none in the United States (as of 2017). Ask students why they think this might be. Possible responses might include the following:

- These trains are too expensive.

- The United States does not have the right kind of tracks.

- The United States is larger than most countries where they are used.

Share with students that there are plans to build maglev trains in some places in the United States, including between San Francisco and Los Angeles, in California, and between New York City and Washington, D.C. You may wish to have students locate these places on a map to highlight the fact that these are short distances relative to the size of the United States.

Students should conclude that maglev trains are a transportation of the future for the United States. Ask students to name what they think the primary means of transportation are now within the United States. Use this information to segue into the Planes, Trains, and Automobiles activity (described in the Activity/Exploration section on p. 121).

Science Class: Students begin to consider maglev train technology in more depth and consider how to apply this understanding to their process of designing their own maglev train prototype for the Maglevacation Train Challenge.

Create a KWL chart and ask students to share what they know already about maglev trains. Then, show one or both of the following videos giving overviews of how maglev trains work (the first gives a very simplistic explanation; the second goes into more depth about electromagnets and polarity and may contain information that your students will not understand at this point):

- *www.discovery.com/tv-shows/other-shows/videos/extreme-engineering-season-1-shorts-maglev-train*

- *www.youtube.com/watch?v=PnDcYgD1VHA*

Continue filling in the KWL chart after watching the videos.

Mathematics Connection: Students explore the importance of budgeting in this activity.

Mathematics Budgeting

Set up 15 to 20 boxes or bags around the room labeled to represent various goods and services, and place a pile of "receipts" in front of each with the name of the item purchased and the cost. Be sure that some items represent wants and others needs, and put them in random order, with wants and needs mixed together. For instance, you might have boxes for the following:

- Housing: $100

- Food: $60

- Haircut or style: $10

- Manicure: $10

- iPhone bill: $20

- Satellite or cable TV: $20

- Car: $100

- Kitten or puppy: $10

- Savings account: $10

- Meal at a restaurant: $10

- New outfit: $20

- Gas for car to get to work: $10

- Shoes (name whatever brand is popular with your students): $20

- Health insurance: $20

- Kindle book: $10

- Trip to the beach: $100

- Gym membership: $20

As students enter the classroom, give them each 20 $10 bills in play money. Tell them that they are now employees of the school, and this is their salary for the week. Give students 5 minutes to go around the room and put money into each box or bag that represents an item or service they wish to "purchase," taking a "receipt" in exchange.

Have students add up their receipts to make sure that they spent their $200 and no more. If any students exceeded $200, they should give up goods or services so that their total is $200. If they are short of $200, they may take additional receipts.

Now, read through the list one item at a time, having students stand up whenever an item they purchased is called out. Save the necessities, such as food, housing, and gas for work, until last. Note whether all students stand up for the necessities, and ask students why they think everyone did or did not spend money on the necessities. Use this activity to introduce the concept of budgeting and making decisions about how to spend money. Tell students that they will create a budget as part of their Maglevacation Train Challenge.

ELA Connection: Tell students that part of their Maglevacation Train Challenge will be to persuade travelers to visit the vacation destination chosen by their design teams. Ask students to brainstorm as a class how they might do this and create a list of their ideas.

Next, show students the following travel video about Paris: *www.youtube.com/watch?v=EDzzsOYPrd4*. Ask students to share their responses about the video's potential to persuade travelers to visit Paris by asking questions such as the following:

- After seeing this video, did you want to visit Paris?

- Why or why not?

- What information did the video give you? (e.g., places to visit, what to do, overall sense that Paris is exciting and fun)

- What information did the video leave out? (e.g., weather, places to stay, cost, how to get there)

- What did the video do to make Paris seem like a good vacation destination?

- What words did the narrator use to persuade the viewer that this is a good place to visit?

Have students make a list of persuasive words that might be used for travel as an entry in their STEM Research Notebooks.

Activity/Exploration

Social Studies Class: This lesson includes two group activities. In Planes, Trains, and Automobiles, students form "expert groups" to research one of several topics that they will compare across three modes of transportation: airplane, train, and car. In Let's Go! each design team work gathers data about their proposed vacation destination. Using the expert group findings along with the design team research, each design team then creates a persuasive presentation, using technology such as PowerPoint if possible, to propose a trip to a fictional client.

Planes, Trains, and Automobiles

Tell students that they are going to have a chance to investigate various modes of travel in this lesson to determine the advantages and disadvantages of the three major means of travel within the United States: airplane, train, and car. Ask students to brainstorm a list of factors travelers might use to decide which method of travel to use (e.g., cost, speed, comfort).

Tell students that they are going to become experts in one of these factors. Divide students into expert groups that each include one member from each design team to enable students to learn using the jigsaw learning method (see Teacher Background Information section on p. 115). The number of groups should thus match the number of students on a design team. See the Explanation section on page 127 for ideas of some expert groups you can create, along with the support you may need to provide each of them.

Once students are in their expert groups, tell them that they are responsible for collecting data about their topic for each of the three types of transportation. They should use the internet to conduct their research and record their findings on the Plains, Trains, and Automobiles graphic organizer (attached at the end of this lesson). Some factors are more subjective than others (e.g., passenger comfort), and groups should be encouraged to provide strong supporting evidence for their decisions.

After the expert groups have finished their research, each should share what it found to be the best and worst modes of transportation and fill it in on a class chart (see Table 4.11 for a sample chart).

Table 4.11. Class Chart for Planes, Trains, and Automobiles Activity

Factor	Mode of Transportation		
	Airplane	Train	Automobile
Cost	*Worst*		*Best*
Availability and convenience			
Environmental factors			
Passenger comfort			
Safety			

Let's Go!

In this activity, the design teams reconvene and act as travel agents working to create a persuasive presentation for a fictional client who wishes to take a four-day trip and has a budget of $2,000 for one person. Each design team conducts research to promote the destination it chose for the Maglevacation Train Challenge and creates a sample budget for its client (the budgeting component of this activity represents the mathematics portion of this lesson).

Teams research various factors about their destination, including climate and entertainment/cultural factors, to include in their presentation to market their destination to the fictional client. Teams also research lodging, cost, and travel options to create a budget for the client using a website such as *www.independenttraveler.com/travel-budget-calculator*. Since there is not currently a maglev train available in the United States, students need to decide on a currently available method of travel—plane, traditional train, or car—and be prepared to support their decision for this choice.

Students then create a narrated presentation, using technology such as PowerPoint if it is available. Each teams' final presentation should include the following elements:

- A persuasive advertising component to highlight the advantages of their destination
- A budget created for their client

Note: If possible, have students print their PowerPoint slides or other presentation materials to include in their Research Notebooks.

Using the EDP to structure student work will help groups move through these tasks in a structured manner. This provides a good opportunity to introduce students to the idea that the EDP can be used to design solutions to problems that do not involve building an actual object. The Let's Go! student handouts (included at the end of this lesson) contain materials to assist students in completing this task.

- Day 1: *Imagine*—Students conduct background research.
- Day 2: *Plan*— Students construct budgets.
- Day 3: *Build*—Students construct work on their presentations.; *Try* (you may wish to assign this task as homework)
- Day 4: *Share*—Students give their presentations.

Science Class: Students learn about electromagnets and relate these to the electromagnets that power maglev trains. They explore propulsion systems and the relationship between a vehicle's weight and its speed in the Balloon Car activity.

Ask students to share what they think causes a maglev train to move. They may believe that magnets cause the train to move, but in fact, the magnets cause the train to levitate, or hover above the tracks. Electricity is used to change the magnets' polarity,

and this switching of poles back and forth propels the train forward. Introduce the term *electromagnets* as devices that create a magnetic field through electricity.

Electromagnetic Demonstration

You will create an electromagnet and have the class predict how the number of coils of wire affects the strength of the magnet. Show students the following materials: a lantern battery, 3 lengths of copper wire, 3 long nails, and 10–15 paper clips. Ask students if any of these materials are magnets. Have students test each material to see if it attracts the paper clips. Explain to students that you can use these materials to create a magnet, but that the magnet will be temporary. Give students the Electromagnetic Demonstration handout (attached at the end of this lesson). Now, do the following as a demonstration (note that the wires may become warm):

- Using the 24-inch piece of copper wire, leave about 6–8 inches of wire loose at one end, and start to wrap the wire around one of the nails, asking students to count how many coils you make around the nail. Leave about 6–8 inches of loose wire on the other side of the nail.

- Show students the other two lengths of copper wire. Ask students to predict whether the amount of wire wrapped around the nail (the number of coils) will influence the strength of the battery.

- Attach the loose ends of the wire that's wrapped around the nail to the battery terminals. Tell students that this is now an electromagnet. Ask students to share their ideas about where the electricity comes from for the electromagnet.

- Remind the students that they need to formulate a testable question for an investigation. Ask students to share ideas about what they will test. Guide students to agree on a testable question such as the following: What happens to the nail's magnetic field with more copper wire wrapped around it? Have students record their question on the Electromagnetic Demonstration handouts in the space provided.

- Ask students how they will measure how strong the battery is. Guide them to the answer that they can measure the strength of the electromagnet by how many paper clips it can pick up. You may wish to create a hook with the first paper clip and hang the other paper clips from it to eliminate the variable of the surface area of the paper clips attracted to the magnet.

- Introduce the concepts of independent variables and dependent variables. Emphasize to students that the *independent variable* is something that is purposely changed during an experiment or test and that a *dependent variable* is the thing that

is measured before and after the independent variable is changed, to see if it has been affected. Have students record the independent and dependent variables on their handouts.

- Have students create a table in the space provided on the student handout to record their predictions and observations for each length of wire and the data for each. If you choose to conduct multiple trials for each, be sure to have students account for this in their data tables.

- Have students sketch the electromagnet in the space provided on the handout, labeling the parts.

- Ask students to predict how many paper clips the electromagnet will attract with the three different lengths of wire and have them record their predictions in the chart they created. Based on these predictions, have students formulate and record a hypothesis. (You may wish to have students formulate this as an "if-then" statement, such as "If we use more wire, then the magnet will be stronger.")

- Suspend the paper clip hook by touching it to the magnet, and have students hang paper clips from the hook until the hook falls away from the magnet. Have students record the data (number of paper clips) in the chart. You may wish to conduct multiple trials for each wire length and have students calculate an average for each trial.

- Repeat for each length of wire, having students count the number of coils for each, and record data.

After the demonstration, have students consider whether their hypothesis was correct and how it was supported or rejected by the data they collected. These ideas, and students' conclusions, should be recorded on the handout.

Remind students that they are going to build a maglev train, but tell them that they will not be using electricity, so they will need to use magnets not only to make their prototype train levitate but also give it a propulsion system, or power to move forward. Ask students to guess what the highest recorded speed of a maglev train is (about 370 mph as of 2016, according to CNN; see *www.cnn.com/videos/world/2015/04/22/ct-japan-maglev-train-world-speed-record.cnn*).

Ask students to brainstorm some ideas about how they could make their prototype train move without touching it (e.g., they could blow on it, use a balloon, attach a magnet to the train and use another magnet to pull or push it). Encourage them to consider ways that do not include magnets. You may wish to use a KWL chart to record student ideas. Tell students that they are going to build cars that are powered by balloons and see what they can learn that might help them in the Maglevacation Train Challenge.

Balloon Car

Have students work together in their design teams or in pairs for this activity. Using the Balloon Car template and following the instructions on page 2 of the Balloon Car handout (attached at the end of this lesson), the teams or pairs each build a balloon car and experiment with the effects of weight on its speed.

First, instruct students in the safe use of scissors. Caution them to always direct the tips of their scissors away from themselves and away from other people. Tell them to be sure to hold the Styrofoam or cardboard behind the scissors as they cut. Also caution students that when working with any sharps, such as pins, toothpicks, or scissors, these things can puncture or cut skin, so they need to be very careful when handling them.

After constructing their balloon cars, have students test their cars and measure the speeds with several different weights. When students have completed the activity, discuss their findings. Topics may include the following:

- What made the car move? (You may wish to discuss Newton's third law—for every action there is an equal and opposite reaction—and ask students to give other examples of this.)

- What happened to the speed as more weights were added? (You may wish to explain to students that when a vehicle is heavier, it accelerates, or increases speed, more slowly, and this is what they observed.)

- Could you make your car heavy enough so that the balloon couldn't cause it to move?

- If this vehicle were a maglev vehicle, could it go faster? Why? (Introduce the concept of friction.)

- What does this mean for the design of your Maglevacation Trains?

Have students reflect on their findings in their STEM Research Notebooks by answering this question: What did you learn from your balloon car that will be helpful in building a maglev train?

Mathematics Connection: The mathematics component of this lesson is integrated within the Let's Go! project.

ELA Connection: The advertising component of the Let's Go! activity is the ELA portion of this lesson. Students should use persuasive language and present relevant and accurate facts in their presentations.

Explanation

This section contains an overview of concepts students should understand in this lesson. You may also wish to have students share what they have learned and explain their understanding of lesson concepts.

Social Studies Class: Students may need support in conducting internet searches for the Let's Go! activity. See the Explanation section in Lesson 2 (p. 94) for more details about conducting internet searches. You may also wish to provide students with summaries of information from the websites suggested below, since some may be significantly above a third-grade reading level.

The expert group topics may require some explanation for students. You may wish to designate the following groups, which may need some support as described:

- Cost analysis group. You may need to support this group in finding a way to compare costs using similar units across modes of transportation. The concept of cost per mile was investigated in Lesson 1 as a mathematics connection (see p. 60); you may wish to review that with this group and guide students to websites that give average costs per mile, such as this one that gives average airfare per mile: *http://airlines.org/data/annual-round-trip-fares-and-fees-domestic*. Alternatively, you may wish to choose a destination and have students research costs of various modes of transportation to reach that destination.

- Availability and convenience group. Students in this group apply their map skills to locate airports, train stations, and auto routes. In particular, you may need to assist students with using map coordinates and indexes to locate train stations, since these often are not marked on maps. Students may wish to start by identifying how available passenger rail stations and airports are in their region. They may extend this by comparing nationwide numbers.

- Environmental factors group. Students in this group should consider pollution by the various forms of transportation. The following article may be helpful: *www.sciencedaily.com/releases/2013/06/130617111345.htm*. Additionally, students may wish to consider impact on wildlife and other environmental factors.

- Passenger comfort group. Many students will not have experienced riding on a plane or train. You may wish to prompt them to think about such things as seat sizes, availability of food, sleeping accommodations, and entertainment options.

- Safety group. These students should consider the availability of seatbelts and other safety features in each form of transportation.

Review EDP procedures with students. Emphasize the importance of collaboration in projects; you may wish to refer to engineers and how they work. Students will be

assessed on collaboration in this lesson (see the Collaboration Rubric attached at the end of this lesson), and they should understand that collaboration requires the following elements and what each one entails:

- Individual accountability: being prepared, communicating with other team members, and completing individual tasks

- Team participation: making a contribution to the team's work, helping solve problems the team encounters, giving feedback to other team members, volunteering to help others when they need help

- Professionalism and respect for team members: being polite and respectful to others and being respectful of other team members' ideas

Mathematics Connection: Students need to understand what a budget is and how to use the columns in the budget handout. Also, you should review how division is used to calculate cost per mile of various forms of transportation (Cost per mile = total cost divided by number of miles traveled) and speed calculations (Speed = distance divided by time).

ELA Connection: Remind students to refer to the list of persuasive words in their Research Notebooks when creating their presentations. Explain to students the importance of providing a reference for information used in a presentation and the importance of tracking this information.

Elaboration/Application of Knowledge

Social Studies Class: Because of the complexity of the Let's Go! activity in this lesson, students have the opportunity to extend and apply their knowledge throughout this lesson, so no additional extensions are provided here.

Science Class: Have students create a STEM Research Notebook entry in response to the prompt below.

STEM Research Notebook Prompt

Students should respond to the following prompt in their Research Notebooks: You created a propulsion system for a vehicle in the Balloon Car activity. Describe what caused the car to move. Think about what the propulsion system for your Maglevacation Train might be. Describe how you would use magnets to cause your train to move. You may include diagrams and sketches.

Mathematics Connection: Not applicable.

ELA Connection: Not applicable.

Evaluation/Assessment

Students may be assessed on the following performance tasks and other measures listed.

Performance Tasks

- Planes, Trains, and Automobiles graphic organizer
- Let's Go! handouts
- Let's Go! presentations (see Presentation Rubric attached at the end of this lesson)
- Electromagnetic Demonstration student handout
- Balloon Car handouts/data sheet
- Persuasive language Research Notebook entry

Other Measures

- Other STEM Research Notebook entries
- Participation in class and group discussions
- Collaboration (see Collaboration Rubric attached at the end of this lesson)

INTERNET RESOURCES

Jigsaw learning method

- *www.jigsaw.org/overview*

Career information

- *www.bls.gov/ooh/sales/travel-agents.htm#tab-2*
- *www.bls.gov/ooh/life-physical-and-social-science/geographers.htm*
- *www.bls.gov/ooh/life-physical-and-social-science/atmospheric-scientists-including-meteorologists.htm#tab-2*

Overview of forms of transportation through history

- *www.geekwire.com/2015/planes-trains-and-automobiles-this-video-is-a-fun-brief-look-at-the-history-of-transportation*

Overview of maglev train technology

- *http://science.howstuffworks.com/transport/engines-equipment/maglev-train.htm*

Discussion of why maglev trains are not currently used in the United States

- *www.cnn.com/2015/05/03/opinions/smart-high-speed-trains-america*

Videos about how maglev trains work

- *www.discovery.com/tv-shows/other-shows/videos/extreme-engineering-season-1-shorts-maglev-train*

- *www.youtube.com/watch?v=PnDcYgD1VHA*

CNN report about the maglev train with the highest recorded speed

- *www.cnn.com/videos/world/2015/04/22/ct-japan-maglev-train-world-speed-record.cnn*

Travel video about Paris

- *www.youtube.com/watch?v=EDzzsOYPrd4*

Average airfare per mile

- *http://airlines.org/data/annual-round-trip-fares-and-fees-domestic*

Travel budget website

- *www.independenttraveler.com/travel-budget-calculator*

Comparison of environmental effects of various forms of transportation

- *www.sciencedaily.com/releases/2013/06/130617111345.htm*

Name: _____

Expert Group: _____

PLANES, TRAINS, AND AUTOMOBILES GRAPHIC ORGANIZER

Record your data and sources below.

Airplane Travel	Train Travel	Automobile (Car) Travel

Based on your research, which of the three modes of transportation is best? Why?

Based on your research, which of the three modes of transportation is worst? Why?

STUDENT HANDOUT, PAGE 1

LET'S GO!

LET'S GO!

You and your design team are challenged to act as travel agents. Your client, Cindy Tourist, has asked you to plan a trip to a fun vacation destination. She has $2,000 to spend on the trip, and it is your task to persuade her that your team's destination is the best spot for her.

Your team will create the following:

- A presentation using PowerPoint or another presentation tool to persuade Cindy Tourist that your team's destination is a great vacation spot
- A budget that shows how she can take this trip for $2,000 or less

Your team will use the EDP to help you work through this challenge.

STUDENT HANDOUT, PAGE 2

LET'S GO!

Use these steps to create your plan for Cindy Tourist's trip:

Define the Problem

Cindy Tourist wants to go on a fun vacation and has $2,000 to spend. Your travel agency wants her to go to the destination your design team chose. You need to persuade her that this is a good place to go and make a plan for her to get there!

Task 1: Learn

Your team first needs to do some research. Use Let's Go! student handout 3 to collect facts about your location. Use your team transportation experts to help you decide how Cindy should travel.

Task 2: Plan

Before you present your ideas to Cindy, you need to be sure that she can afford the trip and the form of transportation you recommend. Use Let's Go! student handout 4 to create a budget for your trip.

Task 3: Try

Create your presentation, using Let's Go! student handout 5.

Task 4: Test

Show your presentation and budget to a family member or other students. Does it convince them that your destination is a good one to visit?

Task 5: Decide

Did you get any ideas about how you could make your presentation better? If so, make changes to improve your presentation and budget.

Task 6

Share your presentation!

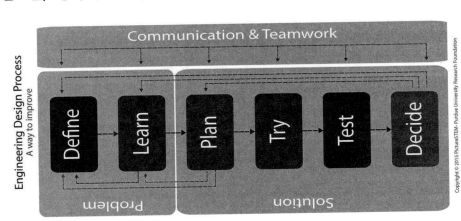

Engineering Design Process
A way to improve

Communication & Teamwork

Define — Learn — Plan — Try — Test — Decide

Problem

Solution

Copyright © 2015 PictureSTEM- Purdue University Research Foundation

Name: _____

Design Team: _____

STUDENT HANDOUT, PAGE 3

LET'S GO!

TASK 1: **LEARN.** WHAT DO YOU NEED TO KNOW? DO SOME RESEARCH!

Our destination is _____.

Find the following information. Be sure to include a reference for where you got the information!

<u>Weather</u> (What is the average temperature? What is the weather like in different seasons?)

<u>Best Way to Get to This Place</u> (Does this place have an airport or a train station? How long would it take to drive there from your location?)

<u>Things to Do</u> (What kind of recreation and entertainment are there? Name at least four.)

<u>Other Reasons People Like to Visit This Spot</u>

Name: _____   Design Team: _____

STUDENT HANDOUT, PAGE 4

LET'S GO!

TASK 2: **PLAN.** MAKE A BUDGET!

Think about everything that Cindy Tourist will need to pay for on her trip.

Use this planner to research travel costs:

Travel Expense	Actual Cost
Transportation (choose one): Air: $ _____ for round-trip ticket Train: $ _____ for round-trip ticket Car: $ _____ for round-trip ticket	
Hotel: 4 nights at $ _____ /night	
Meals: _____ meals at $ _____ /meal	
Entertainment and activities: 1. 2. 3. 4.	
Other expenses (for example, souvenirs):	
Total Cost	

Name: _____   Design Team: _____

STUDENT HANDOUT, PAGE 5

LET'S GO!

TASKS 3–6: **TRY, TEST, DECIDE, AND SHARE.** CREATE YOUR PRESENTATION, TEST IT, MAKE IMPROVEMENTS, AND SHARE IT!

I tested this presentation with _____.

Suggestions for improvement were _____

_____

_____.

Your job is to create a narrated presentation that will persuade Cindy Tourist to visit your team's destination. Your presentation must be between 5 and 7 minutes long. Include the following:

- ✓ Information about your destination's location (city and state)
- ✓ Why Cindy should visit this place
- ✓ What weather she should expect while she is there
- ✓ What there is to do there
- ✓ How she should get there and why this is the best choice
- ✓ Pictures of the destination and things to do
- ✓ Narration that uses persuasive language
- ✓ All team members' participation in the narration
- ✓ A budget that demonstrates she can take this trip for $2,000 or less

An excellent presentation will include the following:

Information	• Team includes interesting and factual information about the destination for all categories (location, transportation, weather, and entertainment/recreation).
Ideas and organization	• Team has a clear main idea and organizational strategy. • Presentation includes both an introduction and conclusion. • Presentation is coherent, well organized, and informative. • Team uses presentation time well, and presentation is between 5 and 7 minutes long.
Presentation style	• All team members participate in the presentation. • Presenters are easy to understand. • Presenters use appropriate language for audience (no slang or poor grammar, infrequent use of filler words such as "uh," "um").
Visual aids	• Team uses well-produced visual aids or media that clarify and enhance presentation.
Response to questions	• Team responds clearly and in detail to audience questions and seeks clarification of questions.

Name: _____   Design Team: _____

STUDENT HANDOUT, PAGE 1

BALLOON CAR

You will create a car that uses a balloon to propel it forward. You will test the car with different amounts of weight.

The testable question is _____

_____.

The independent variable is _____

The dependent variable is _____.

My hypothesis is _____

_____

_____

_____.

STUDENT HANDOUT, PAGE 2

BALLOON CAR

Safety note: Safety glasses or goggles are required for this activity.

Instructions

1. Using the templates, trace the shapes onto the Styrofoam or cardboard.

2. Choose one person who will blow up the balloon throughout the activity.

3. Insert the bendy end of the straw a short distance into the balloon and wrap tape around it to seal it to the straw.

4. Tape your straw in the middle of the rectangle as shown, with the balloon hanging off the back

5. Attach your wheels: Carefully push one of the pins through the center of the smaller circle, then into the center of the larger circle, then into the edge of the rectangular piece. (Make sure to leave some space between the rectangle and the wheels so that the wheels can turn.) Attach all 4 wheels this way.

6. Take a test run. Use the straw to blow up the balloon, then let it go.

7. Time trials: Pick a starting point on the floor, and use a piece of masking tape to mark the starting line. Measure 1 meter from the starting line, and put another piece of masking tape there to mark the finish line.

8. Choose a timekeeper and a starter. (Switch for the different trials.) Remember, the same person will always blow up the balloon.

9. The starter says, "Go!" when the car is ready to go. This is the signal for the balloon blower to release the car and the timekeeper to start timing. The timekeeper uses the stopwatch or timer to measure how long it takes the car to go from the start to the finish line. When the car crosses the finish line, note how much time it took in the chart on page 3 of the handout.

10. Now, put 5 weights on the car and repeat for 3, 7, and 10 weights, recording your times in the chart.

Name: _____

STUDENT HANDOUT, PAGE 3

BALLOON CAR

Weights	Time to Travel 1 Meter	Speed (Seconds per Meter)
0		
3		
5		
7		
10		

What did the data tell you about your hypothesis? _____

_____

_____

_____

_____

My conclusions: _____

_____

_____

_____

_____

BALLOON CAR TEMPLATES

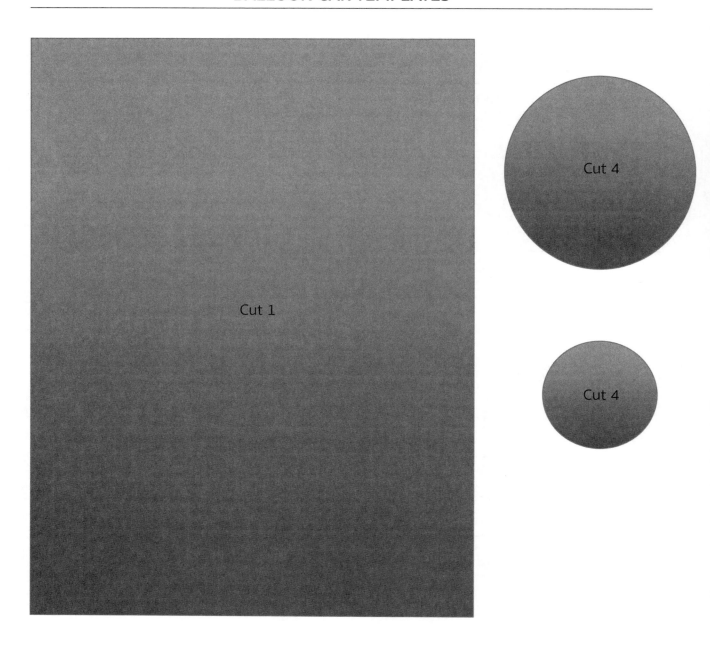

Cut 1

Cut 4

Cut 4

NATIONAL SCIENCE TEACHERS ASSOCIATION

Team Name: _____

Let's Go! Presentation Rubric (30 points possible)

Team Performance	Below Standard (0–2)	Approaching Standard (3–4)	Meets or Exceeds Standard (5–6)	Team Score
INFORMATION	• Team includes little interesting or factual information about the destination and may omit categories (location, transportation, weather, and entertainment/recreation).	• Team includes some interesting and factual information about the destination for most but not all categories (location, transportation, weather, and entertainment/recreation).	• Team includes interesting and factual information about the destination for all categories (location, transportation, weather, and entertainment/recreation).	
IDEAS AND ORGANIZATION	• Team does not have a main idea or organizational strategy. • Presentation does not include an introduction and/or conclusion. • Presentation is confusing and uninformative. • Team uses presentation time poorly, and it is too short or too long (less than 5 minutes or longer than 7 minutes).	• Team has a main idea and organizational strategy, although it may not be clear. • Presentation includes either an introduction or conclusion or both. • Presentation is fairly coherent, well organized, and informative. • Team uses presentation time adequately, but presentation may be slightly too short or too long (less than 5 minutes or longer than 7 minutes).	• Team has a clear main idea and organizational strategy. • Presentation includes both an introduction and conclusion. • Presentation is coherent, well organized, and informative. • Team uses presentation time well, and presentation is between 5 and 7 minutes long.	
PRESENTATION STYLE	• Only one or two team members participate in the presentation. • Presenters are difficult to understand. • Presenters use language inappropriate for audience (slang, poor grammar, frequent filler words such as "uh," "um").	• Some, but not all, team members participate in the presentation. • Most presenters are understandable, but volume may be too low or some presenters may mumble. • Presenters use some language inappropriate for audience (slang or poor grammar, some use of filler words such as "uh," "um").	• All team members participate in the presentation. • Presenters are easy to understand. • Presenters use appropriate language for audience (no slang or poor grammar, infrequent use of filler words such as "uh," "um").	

(continued)

Team Performance	Below Standard (0–2)	Approaching Standard (3–4)	Meets or Exceeds Standard (5–6)	Team Score
VISUAL AIDS	• Team does not use any visual aids in presentation. • Visual aids are used but do not add to the presentation.	• Team uses some visual aids in presentation, but they may be poorly executed or distract from the presentation.	• Team uses well-produced visual aids or media that clarify and enhance presentation.	
RESPONSE TO AUDIENCE QUESTIONS	• Team fails to respond to questions from audience or responds inappropriately.	• Team responds appropriately to audience questions, but responses may be brief, incomplete, or unclear.	• Team responds clearly and in detail to audience questions and seeks clarification of questions.	

TOTAL SCORE: _____

Let's Go! Collaboration Rubric (30 points possible)

Student Name: _____   Team Name: _____

Individual Performance	Below Standard (0–3)	Approaching Standard (4–7)	Meets or Exceeds Standard (8–10)	Student Score
INDIVIDUAL ACCOUNTABILITY	• Student is unprepared. • Student does not communicate with team members and does not manage tasks as agreed on by the team. • Student does not complete or participate in project tasks. • Student does not complete tasks on time. • Student does not use feedback from others to improve work.	• Student is usually prepared. • Student sometimes communicates with team members and manages tasks as agreed on by the team, but not consistently. • Student completes or participates in some project tasks but needs to be reminded. • Student completes most tasks on time. • Student sometimes uses feedback from others to improve work.	• Student is consistently prepared. • Student consistently communicates with team members and manage tasks as agreed on by the team. Student discusses and reflects on ideas with the team. • Student completes or participates in project tasks without being reminded. • Student completes tasks on time. • Student uses feedback from others to improve work.	
TEAM PARTICIPATION	• Student does not help the team solve problems; may interfere with teamwork. • Student does not express ideas clearly, pose relevant questions, or participate in group discussions. • Student does not give useful feedback to other team members. • Student does not volunteer to help others when needed.	• Student cooperates with the team but may not actively help solve problems. • Student sometimes expresses ideas, poses relevant questions, elaborates in response to questions, and participates in group discussions. • Student provides some feedback to team members. • Student sometimes volunteers to help others.	• Student helps the team solve problems and manage conflicts. • Student makes discussions effective by clearly expressing ideas, posing questions, responding thoughtfully to team members' questions and perspectives. • Student gives useful feedback to others so they can improve their work. • Student volunteers to help others if needed.	

(continued)

Individual Performance	Below Standard (0–3)	Approaching Standard (4–7)	Meets or Exceeds Standard (8–10)	Student Score
PROFESSIONALISM AND RESPECT FOR TEAM MEMBERS	• Student is impolite or disrespectful to other team members. • Student does not acknowledge or respect others' ideas and perspectives.	• Student is usually polite and respectful to other team members. • Student usually acknowledges and respects others' ideas and perspectives.	• Student is consistently polite and respectful to other team members. • Student consistently acknowledges and respects others' ideas and perspectives.	

TOTAL SCORE: _____

Lesson Plan 4: Speeding Ahead—The Maglevacation Train Challenge

In this lesson, students apply their knowledge of mapping, geography, trains, and science concepts related to magnetism and speed to design their Maglevacation Trains. The challenge draws on materials that they have produced in previous lessons. In particular, the Let's Go! activity destination information from Lesson 3 provides the persuasive destination content for this challenge. Each student team will design and build its train and prepare a video presentation that demonstrates the train, highlights its design features and advantages, demonstrates students' understanding of magnetism, and markets the team's vacation destination to clients. You may wish to enlist classroom visitors, such as school administrators, local travel agents, engineers, or other students, to act as the audience for these presentations.

ESSENTIAL QUESTION

How can we combine our understanding of mapping, geography, trains, and magnetism to design a solution to the Maglevacation Train Challenge?

ESTABLISHED GOALS AND OBJECTIVES

At the conclusion of this lesson, students will be able to do the following:

- Apply their understanding of the EDP to work collaboratively to create a solution to a challenge

- Apply their understanding of mapping and geography to create a persuasive presentation about a travel destination

- Apply their understanding of magnetism to create a prototype maglev train

- Apply their understanding of speed and how weight affects acceleration to achieve the fastest speeds possible over a short distance

- Apply their understanding of the EDP to design and build a prototype train

- Synthesize what they learned throughout the module to create a presentation appropriate for their audience

TIME REQUIRED

- 8 days (approximately 45 minutes each day; see Tables 3.9–3.10)

MATERIALS

Required Materials for Lesson 4

- STEM Research Notebooks
- Access to internet technology for showing video clips and for student research
- Classroom map of the United States
- Chart paper
- Handouts (attached at the end of this lesson)

Additional Materials for Maglevacation Train Challenge (per team unless otherwise indicated)

- Safety glasses or goggles (1 pair per student)
- Equipment to film student videos
- 1 piece of lightweight cardboard
- 3 bar magnets
- 4 small, round magnets
- 1 roll masking tape
- 5 pennies
- 4 paper clips
- 6 toothpicks
- 2 sheets of construction paper
- 2 plastic figures (the "passengers")
- 2 straws
- 2 balloons
- Scissors
- Ruler
- Markers
- Maglevacation Train Challenge student packet (1 per student)
- 6 magnetic ceramic bars, each 1 ½ inches wide x 3 inches long (per class)
- 2 corrugated cardboard pieces each 1 meter long for track (per class)
- Duct tape for track (per class)

CONTENT STANDARDS AND KEY VOCABULARY

Table 4.12 lists the content standards from the *NGSS*, the *Common Core State Standards*, and the Framework for 21st Century Learning that this lesson addresses, and Table 4.13 (p. 151) presents the key vocabulary. Vocabulary terms are provided for both teacher and student use. Teachers may choose to introduce some or all of the terms to students.

Table 4.12. Content Standards Addressed in STEM Road Map Module Lesson 4

NEXT GENERATION SCIENCE STANDARDS

PERFORMANCE EXPECTATIONS

- 3-PS2-3. Ask questions to determine cause and effect relationships of electric or magnetic interactions between two objects not in contact with each other.

- 3-PS2-4. Define a simple design problem that can be solved by applying scientific ideas about magnets.

- 3-5-ETS1-1. Define a simple design problem reflecting a need or a want that includes specified criteria for success and constraints on materials, time, or cost.

- 3-5-ETS1-2. Generate and compare multiple possible solutions to a problem based on how well each is likely to meet the criteria and constraints of the problem.

- 3-5-ETS1-3. Plan and carry out fair tests in which variables are controlled and failure points are considered to identify aspects of a model or prototype that can be improved.

DISCIPLINARY CORE IDEAS

PS2.A: Forces and Motion

- Each force acts on one particular object and has both strength and a direction. An object at rest typically has multiple forces acting on it, but they add to give zero net force on the object. Forces that do not sum to zero can cause changes in the object's speed or direction of motion.

PS2.B: Types of Interactions

- Electric and magnetic forces between a pair of objects do not require that the objects be in contact. The sizes of the forces in each situation depend on the properties of the objects and their distances apart and, for forces between two magnets, on their orientation relative to each other.

ETS1.A: Defining and Delimiting Engineering Problems

- Possible solutions to a problem are limited by available materials and resources (constraints). The success of a designed solution is determined by considering the desired features of a solution (criteria). Different proposals for solutions can be compared on the basis of how well each one meets the specified criteria for success or how well each takes the constraints into account. (3-5-ETS1-1)

ETS1.B: Developing Possible Solutions

- Research on a problem should be carried out before beginning to design a solution. Testing a solution involves investigating how well it performs under a range of likely conditions. (3-5-ETS1-2)

- At whatever stage, communicating with peers about proposed solutions is an important part of the design process, and shared ideas can lead to improved designs. (3-5-ETS1-2)

- Tests are often designed to identify failure points or difficulties, which suggest the elements of the design that need to be improved. (3-5-ETS1-3)

ETS1.C: Optimizing the Design Solution

- Different solutions need to be tested in order to determine which of them best solves the problem, given the criteria and the constraints. (3-5-ETS1-3)

CROSSCUTTING CONCEPTS

Scale, Proportion, and Quantity

- Standard units are used to measure and describe physical quantities such as weight, time, temperature, and volume.

Energy and Matter

- Energy can be transferred in various ways and between objects.

Cause and Effect

- Cause and effect relationships are routinely identified, tested, and used to explain change.

Influence of Science, Engineering, and Technology on Society and the Natural World

- People's needs and wants change over time, as do their demands for new and improved technologies.

- Engineers improve existing technologies or develop new ones to increase their benefits, decrease known risks, and meet societal demands.

SCIENCE AND ENGINEERING PRACTICES

Asking Questions and Defining Problems

- Ask questions about what would happen if a variable is changed.

- Use prior knowledge to describe problems that can be solved.

- Define a simple design problem that can be solved through the development of an object, tool, process, or system and includes several criteria for success and constraints on materials, time, or cost.

Engaging in Argument From Evidence

- Compare and refine arguments based on an evaluation of the evidence presented.

- Respectfully provide and receive critiques from peers about a proposed procedure, explanation, or model by citing relevant evidence and posing specific questions.

- Construct and/or support an argument with evidence, data, and/or a model.

- Use data to evaluate claims about cause and effect.

- Make a claim about the merit of a solution to a problem by citing relevant evidence about how it meets the criteria and constraints of the problem.

Obtaining, Evaluating, and Communicating Information

- Combine information in written text with that contained in corresponding tables, diagrams, and/or charts to support the engagement in other scientific and/or engineering practices.

- Obtain and combine information from books and/or other reliable media to explain phenomena or solutions to a design problem.

- Communicate scientific and/or technical information orally and/or in written formats, including various forms of media as well as tables, diagrams, and charts.

Developing and Using Models

- Identify limitations of models.

- Collaboratively develop and/or revise a model based on evidence that shows the relationships among variables for frequent and regular occurring events.

- Develop a diagram or simple physical prototype to convey a proposed object, tool, or process.

- Use a model to test cause and effect relationships or interactions concerning the functioning of a natural or designed system.

Constructing Explanations and Designing Solutions

- Use evidence (e.g., measurements, observations, patterns) to construct or support an explanation or design a solution to a problem.

- Identify the evidence that supports particular points in an explanation.

- Apply scientific ideas to solve design problems.

- Generate and compare multiple solutions to a problem based on how well they meet the criteria and constraints of the design solution.

Using Mathematical and Computational Thinking

- Describe, measure, estimate, and/or graph quantities (e.g., area, volume, weight, time) to address scientific and engineering questions and problems.

- Organize simple data sets to reveal patterns that suggest relationships.

COMMON CORE STATE STANDARDS FOR MATHEMATICS

MATHEMATICAL PRACTICES

- MP1. Make sense of problems and persevere in solving them.

- MP2. Reason abstractly and quantitatively.

- MP4. Model with mathematics.
- MP5. Use appropriate tools strategically.
- MP6. Attend to precision.

MATHEMATICAL CONTENT

- NBT.A.2. Fluently add and subtract within 1000 using strategies and algorithms based on place value, properties of operations, and/or the relationship between addition and subtraction.
- 3.MD.A.1. Tell and write time to the nearest minute and measure time intervals in minutes. Solve word problems involving addition and subtraction of time intervals in minutes, e.g., by representing the problem on a number line diagram.

COMMON CORE STATE STANDARDS FOR ENGLISH LANGUAGE ARTS

WRITING STANDARDS

- W.3.1. Write opinion pieces on topics or texts, supporting a point of view with reasons.
- W.3.1.A. Introduce the topic or text they are writing about, state an opinion, and create an organizational structure that lists reasons.
- W.3.1.B. Provide reasons that support the opinion.
- W.3.1.C. Use linking words and phrases (e.g., *because, therefore, since, for example*) to connect opinion and reasons.
- W.3.8. Recall information from experiences or gather information from print and digital sources; take brief notes on sources and sort evidence into provided categories.

SPEAKING AND LISTENING STANDARDS

- SL.3.1. Engage effectively in a range of collaborative discussions (one-on-one, in groups, and teacher-led) with diverse partners on *grade 3 topics and texts,* building on others' ideas and expressing their own clearly.
- SL3.1.D. Explain their ideas and understanding in light of the discussion.
- SL3.3. Ask and answer questions about information from a speaker, offering appropriate elaboration and detail.
- SL3.4. Report on a topic or text, tell a story, or recount an experience with appropriate facts and relevant, descriptive details, speaking clearly at an understandable pace.
- SL.3.6. Speak in complete sentences when appropriate to task and situation in order to provide requested detail or clarification.

FRAMEWORK FOR 21ST CENTURY LEARNING

Interdisciplinary themes (inventions, history, engineering design process, and progress); Learning and Innovation Skills; Information, Media and Technology Skills; Life and Career Skills

Table 4.13. Key Vocabulary in Lesson 4

Key Vocabulary	Definition
feedback	information about someone's reaction to a product or presentation that can be used to improve that product or presentation

TEACHER BACKGROUND INFORMATION

Students apply what they learned throughout the module in this lesson. You may wish to review the EDP with students prior to the start of the challenge, since they will rely on this process for the design and build portion. By this point in the module, students should have access to a large amount of information in their STEM Research Notebooks. These Research Notebooks will be key resources for students as they approach their final challenge. In particular, the findings from the Let's Go! activity in Lesson 3 will form the basis for part of their videos.

You may wish to invite outside guests into the classroom to view student videos and give feedback. It is important to provide a structured way for these individuals to give students feedback. Providing guests with a simple form such as the following gives them a structured way to share their thoughts with students:

Design Team Name: _____ Destination: _____

What I really liked about your presentation was:

I was curious about:

You could improve on:

I have / have not (circle one) been to your destination.

COMMON MISCONCEPTIONS

Students will have various types of prior knowledge about the concepts introduced in this lesson. Table 4.14 (p. 152) outlines some common misconceptions students may have concerning these concepts. Because of the breadth of students' experiences, it is not possible to anticipate every misconception that students may bring as they approach this lesson. Incorrect or inaccurate prior understanding of concepts can influence student

learning in the future, however, so it is important to be alert to misconceptions such as those presented in the table.

Table 4.14. Common Misconceptions About the Concepts in Lesson 4

Topic	Student Misconception	Explanation
Forces	Nonmoving objects do not exert a force.	Stationary objects can exert forces on other objects. For example, when you roll a toy car over a tabletop, there is a frictional force between the nonmoving tabletop and the car wheels.
	If a moving object is slowing, the force that was propelling it forward is decreasing.	When a force acts on a moving object in a direction opposite the direction of motion, the moving object will slow, even if the force that was propelling the object forward continues.
Magnetism	All metals are magnetic.	The only naturally occurring magnetic metals are iron, cobalt, and nickel.
	All magnets are solids.	Magnetic fields can be created in space by electric currents.
	Large magnets exert a stronger magnetic field than small magnets.	The size of a magnet and its magnetism are not necessarily related. The substances that compose the magnet, not its size, determine its magnetic force.

PREPARATION FOR LESSON 4

Review the Teacher Background Information provided, assemble the materials for the lesson, and preview the videos recommended in the Learning Plan Components section below. Students will be assessed on collaboration and each design team will be assessed on its presentation and train design as part of this lesson. You may wish to review the associated rubrics (attached at the end of this lesson).

This lesson requires some specific preparation. You need to ensure that students have access to the appropriate technology to create their videos. Also, you need to create a track for the students' trains for testing and demonstration purposes. This can be done fairly simply with strip magnets and pieces of cardboard. Your track will need to be at least 1 meter long. Place the ceramic magnets end to end on a tabletop to create a track that is about 1 meter long. The track should be about 6 inches wide with the ceramic magnets at approximately the center of the track. Use corrugated cardboard pieces 1 meter long ×

4–6 inches high to create walls along each side of the length of the track. Use duct tape to secure the walls to the table, keeping the walls as close to the magnet strips as possible.

LEARNING PLAN COMPONENTS
Introductory Activity/Engagement

Connection to the Challenge: In this lesson, students actively address the driving question for the module and challenge: How can we create a plan and build a prototype for a maglev train to carry passengers to a vacation destination? Begin each day by having students summarize their work from previous days' lessons and describe how their work contributed to answering the driving question and creating a solution for the module challenge.

Social Studies Class: Show students a Jetsons video featuring a 1960s view of future technology (visit YouTube and search for "The Jetsons Future of Technology" or access the video directly at *www.youtube.com/watch?v=e8SC6bny1SA*). Point out to students that this show was created in the 1960s, and the future it was envisioning is now. Create a class list of what technologies the video predicted actually exist and what technologies do not exist (or are not commonly used).

Tell students that they've reached the final stage of their challenge now, and they will be thinking to the future. Ask students to recall what the challenge is (the Maglevacation Train Challenge) and what they need to do. Remind students that there are currently no maglev trains in the United States, so they are proposing something innovative. Tell students that in this lesson, they will receive all the design parameters or rules for the challenge as well as their supplies.

Describe the challenge to students: Families in your town want a quick and easy way to travel to a vacation spot. Since you are learning about magnetic levitation trains, your town transportation department has asked you to design a prototype for a train that can take families quickly and safely to a vacation destination. You and your team will act as the travel agents and the train design engineers. Your team chose a destination for your train, now you and your team will create a prototype Maglevacation Train, and provide information about the destination and train to your clients. All aboard!

Science Class: Tell students that trains need a way of being propelled forward. Maglev trains use electrified coils in the track and the walls along the tracks to move the train as it levitates. Since students will not have access to electromagnets in this lesson, each team will choose a method of propulsion to move its Maglevacation Trains along the track. They may use a balloon or another method of propulsion if they can create it with the supplies provided. Have students brainstorm about what might be the fastest method of propulsion and offer explanations for why they believe that.

Ask students to brainstorm about what methods of propulsion have been used for trains over the years. Remind them to consider wagon tramways, steam engines, and

other types of trains they may know about. Create a class list of train propulsion systems, and ask students to put these in order from earliest to most recent.

Have student groups research modern trains to learn how they are powered. Hold a class discussion about technological innovation and what effect this has had on train travel (e.g., faster, safer, more environmentally friendly). You may wish to use or duplicate the timeline created in Lesson 2 as a reminder about types of trains and add technological innovations to this timeline.

Mathematics Connection: Tell students that each team is going to design its Maglevacation Train so that it travels as fast as possible. Ask students what the measure is for how fast something goes (speed). Remind them that this is expressed as distance/time. Show the following video featuring the fastest vehicles in the world: *www.youtube.com/watch?v=SsDZ7-esM_4*. Then, have students predict how fast they can get their prototype train to go (in meters/second) and record those predictions in their STEM Research Notebooks.

ELA Connection: Ask students if they have ever watched someone reporting the news on TV or seen someone speaking in public. Ask what makes that person interesting to listen to and what keeps the audience's attention. Show a sample local news broadcast if you can find one that is age-appropriate for your students. Then, have students create a list of what the newscaster did that made him or her interesting and easy to listen to.

Next, remind students that each team will be recorded on a video as it present its Maglevacation Train Challenge projects. Tell them that learning to speak publicly, just like other kinds of learning, requires practice. Pair students and allow them to interview one another about their favorite vacation spots. Give students 3 minutes to interview each other, and then have them use the presentation skills they identified as each student introduces his or her partner's favorite vacation spot to the class in 30 seconds.

Activity/Exploration

Because this final challenge incorporates students' learning from all subject areas in the module, this section is presented as a truly integrated project and is separated into subject areas. The tasks associated with the challenge can, however, be completed within various classes if you wish.

The Maglevacation Train Challenge

In this challenge, students teams each create a comprehensive plan for their vacation destination and their train. The Maglevacation Train Challenge student packet (attached at the end of this lesson) provides details on how students should work through this challenge. The background work for some parts of the plan were completed in previous lessons (denoted by an asterisk in the list below). Student teams need to draw on this

prior work and students' accumulated knowledge from throughout the unit to produce team video presentations of 10 minutes or less that include the following elements:

- An introduction to their team

- An introduction to their destination, including geographic facts*

- Informative materials about their destination*

- A demonstration of their maglev train

- Information about their train's speed

- A discussion of the features of their train that makes it unique

- A persuasive argument as to why travelers should visit the chosen vacation destination*

- A persuasive argument as to why travelers should choose to use the Maglevacation Train

On the final day of the module, student teams will each present their videos to the class and guests. These guests may include school administrators, parents, and perhaps local travel agents or engineers. After the team presentations, the audience should be allowed the opportunity to ask questions.

Explanation

This section contains an overview of concepts students should understand in this lesson. You may also wish to have students share what they have learned and explain their understanding of lesson concepts.

Science Class/Social Studies Class: Be sure that students are aware of the challenge rules (see student packet attached at the end of this lesson). Explain to students that they will be assessed on collaboration, their team's maglev train design, and their team's presentation (see rubrics attached at the end of this lesson).

Mathematics Connection: Not applicable.

ELA Connection: Not applicable.

Elaboration/Application of Knowledge

Social Studies Class: Hold a class discussion about the implications of various forms of transportation in the future. Topics might include resource scarcity and fossil fuels, environmental concerns, and improving technology. Then, have students create a STEM Research Notebook entry in response to the prompt below.

STEM Research Notebook Prompt

Students should respond to the following prompt in their Research Notebooks: *Do you think that high-speed trains would be useful in the United States? Why or why not? Why do you think we do not have these types of trains in the United States currently?*

Science Class: Students can reflect on their maglev train designs and how their trains used the science concepts they learned during this module.

Mathematics Connection: Have students conduct multiple trials with their maglev train prototypes. Using these data, students can calculate average speed and create charts of graphs of the various speeds of design teams' trains.

ELA Connection: Have each student write a paragraph in their STEM Research Notebooks describing a mode of transportation that might be used in the year 2050.

Evaluation/Assessment

Students may be assessed on the following performance tasks and other measures listed.

Performance Tasks

- Maglevacation Train Challenge student packet:
  - Design Time! handouts
  - Lights, Camera, Action handouts
- Prototype design (see Prototype Design Rubric attached at the end of this lesson)
- Video presentation (see Presentation Rubric attached at the end of this lesson)

Other Measures

- STEM Research Notebook entries
- Participation in class and group discussions
- Collaboration (see Collaboration Rubric attached at the end of this lesson)

INTERNET RESOURCES

Jetsons video featuring a 1960s view of future technology

- *www.youtube.com/watch?v=e8SC6bny1SA*

Video featuring the fastest vehicles in the world

- *www.youtube.com/watch?v=SsDZ7-esM_4*

4

STUDENT PACKET, PAGE 1

THE MAGLEVACATION TRAIN CHALLENGE

Families in your town want a quick and easy way to travel to a vacation spot. Since you are learning about magnetic levitation trains, your town transportation department has asked you to design a prototype for a train that can take families quickly and safely to a vacation destination.

You and your team will act as the travel agents and the train design engineers. You will choose a destination for your train and create a prototype Maglevacation Train.

All aboard!

STUDENT PACKET, PAGE 2

WELCOME TO THE MAGLEVACATION TRAIN CHALLENGE!

You and your team will have the chance to use your knowledge about geography, trains, magnets, and engineering design in this challenge. You will create the following:

1. A prototype Maglevacation Train

2. A video about your vacation destination and your train (8–10 minutes). You will be assessed on the following:

   ✓ Collaboration (how well you work with your team)

   ✓ Your train design—how it works and how it looks (it should be visually appealing)

   ✓ The information you provide in your video about your train's design, features, and speed

   ✓ The information you provide in your video about your destination and how to get there

   ✓ The overall content and persuasiveness of your video (feel free to be creative!)

Rules for the Challenge

   ✓ Everyone on the team must participate in the train design and in the video.

   ✓ You may use only the materials provided to build your train. You do not need to use them all!

   ✓ You must assess safety issues and take appropriate safety precautions.

   ✓ You must use the EDP to design and build your train.

   ✓ Your train must be between 3 and 5 inches wide and at least 3 inches long.

   ✓ You may not touch your train with your hands to make it move. (You may push or pull it along the track with magnets.)

   ✓ Your train should be able to travel 1 meter on the track provided.

   ✓ Your train must carry your "passengers" safely to the end of the track.

   ✓ You can use any information or resources from previous lessons. Use your Research Notebooks!)

4

Name: _____   Team Name: _____

STUDENT PACKET, PAGE 3

DESIGN TIME!
MAGLEVACATION TRAIN CHALLENGE

Step 1: (Define)   State the problem (what are you trying to do?):

Step 2: (Learn)   What solutions can you and your team imagine?

Step 3: (Plan)   Make a sketch of your design here and label materials (use the back of the page if you need more space). Be sure to consider safety issues when planning your design.

Step 4: (Try)   Build it! Did everyone on your team have a chance to help?

DESIGN TIME!
MAGLEVACATION TRAIN CHALLENGE

Step 5: (Test)   Test your design. How did it work? Could it work better?

What is your train's best speed? _____

Step 6: (Decide)   This is your chance to use your test results to decide what changes
you should make. What did you change?

Step 7: Share your design! What design features of your train do you want to highlight
in your video? Be sure to think about what makes your train unique, safe, and
appealing to passengers.

In your video, you should discuss what you learned and what science concepts were useful
to you in responding to this challenge.

Name: _____   Team Name: _____

STUDENT PACKET, PAGE 5

LIGHTS, CAMERA, ACTION! MAGLEVACATION TRAIN CHALLENGE

Every television show, commercial, or professionally produced video requires planning and organization. Your team is going to plan the elements of your video using this graphic organizer.

Introduction of Your Team

Presentation features to include in this portion of your video (for example, music or images):

Team information to include:

Who will present this introduction?

Description of Your Destination

Presentation features to include in this portion of your video (for example, music or images):

Destination location (be sure to show the location on a map):

Distance from home:

Weather:

Have any of your team members been to this place before?

Who will present this description?

Name: _____   Team Name: _____

STUDENT PACKET, PAGE 6

LIGHTS, CAMERA, ACTION!
MAGLEVACATION TRAIN CHALLENGE

Information About Your Destination

Presentation features to include in this portion of your video (for example, music or images):

What is there to do in this place?

Why do people like to go there?

Is there anything else interesting about this place?

Who will present the destination information?

Information About Your Maglevacation Train

Presentation features to include in this portion of your video (for example, music or images):

Who will demonstrate your train on the track?

What is your train's top speed?

What makes your train special and a good way to travel?

Who will present the train information?

NATIONAL SCIENCE TEACHERS ASSOCIATION

4

Name: _____   Team Name: _____

STUDENT PACKET, PAGE 7

LIGHTS, CAMERA, ACTION! MAGLEVACATION TRAIN CHALLENGE

Persuasive Information

Presentation features to include in this portion of your video (for example, music or images):

Anything Else You Want to Include!

Why should travelers choose your vacation destination?

Why should they take your train to get there?

How will you conclude your video?

Who will present this persuasive information?

Transportation in the Future Lesson Plans

Maglevacation Train Presentation Rubric (30 points possible)

Team Name: _____

Team Performance	Below Standard (0–2)	Approaching Standard (3–4)	Meets or Exceeds Standard (5–6)	Team Score
INFORMATION	• Team includes little interesting and factual information about the destination and may omit categories (location, transportation, weather, and entertainment/recreation).	• Team includes some interesting and factual information about the destination for most but not all categories (location, transportation, weather, and entertainment/ recreation).	• Team includes interesting and factual information about the destination for all categories (location, transportation, weather, and entertainment/recreation).	
IDEAS AND ORGANIZATION	• Team does not have a main idea or organizational strategy. • Presentation does not include an introduction and/or conclusion. • Presentation is confusing and uninformative. • Team uses presentation time poorly, and it is too short or too long (less than 8 minutes or longer than 10 minutes).	• Team has a main idea and organizational strategy, although it may not be clear. • Presentation includes either an introduction or conclusion or both. • Presentation is fairly coherent, well organized, and informative. • Team uses presentation time adequately, but presentation may be slightly too short or too long (less than 8 minutes or longer than 10 minutes).	• Team has a clear main idea and organizational strategy. • Presentation includes both an introduction and conclusion. • Presentation is coherent, well organized, and informative. • Team uses presentation time well, and presentation is between 8 and 10 minutes long.	
PRESENTATION STYLE	• Only one or two team members participate in the presentation. • Presenters are difficult to understand. • Presenters use language inappropriate for audience (slang, poor grammar, frequent filler words such as "uh," "um").	• Some, but not all, team members participate in the presentation. • Most presenters are understandable, but volume may be too low or some presenters may mumble. • Presenters use some language inappropriate for audience (slang or poor grammar, some use of filler words such as "uh," "um").	• All team members participate in the presentation. • Presenters are easy to understand. • Presenters use appropriate language for audience (no slang or poor grammar, infrequent use of filler words such as "uh," "um").	

Team Performance	Below Standard (0–2)	Approaching Standard (3–4)	Meets or Exceeds Standard (5–6)	Team Score
VISUAL AIDS	• Team does not use any visual aids in presentation. • Visual aids are used but do not add to the presentation.	• Team uses some visual aids in presentation, but they may be poorly executed or distract from the presentation.	• Team uses well-produced visual aids or media that clarify and enhance presentation.	
RESPONSE TO AUDIENCE QUESTIONS	• Team fails to respond to questions from audience or responds inappropriately.	• Team responds appropriately to audience questions, but responses may be brief, incomplete, or unclear.	• Team responds clearly and in detail to audience questions and seeks clarification of questions.	
TOTAL SCORE: _____				

Maglevacation Train Collaboration Rubric (30 points possible)

Student Name: _____   Team Name: _____

Individual Performance	Below Standard (0–3)	Approaching Standard (4–7)	Meets or Exceeds Standard (8–10)	Student Score
INDIVIDUAL ACCOUNTABILITY	• Student is unprepared. • Student does not communicate with team members and does not manage tasks as agreed on by the team. • Student does not complete or participate in project tasks. • Student does not complete tasks on time. • Student does not use feedback from others to improve work.	• Student is usually prepared. • Student sometimes communicates with team members and manages tasks as agreed on by the team, but not consistently. • Student completes or participates in some project tasks but needs to be reminded. • Student completes most tasks on time. • Student sometimes uses feedback from others to improve work.	• Student is consistently prepared. • Student consistently communicates with team members and manage tasks as agreed on by the team. Student discusses and reflects on ideas with the team. • Student completes or participates in project tasks without being reminded. • Student completes tasks on time. • Student uses feedback from others to improve work.	
TEAM PARTICIPATION	• Student does not help the team solve problems; may interfere with teamwork. • Student does not express ideas clearly, pose relevant questions, or participate in group discussions. • Student does not give useful feedback to other team members. • Student does not volunteer to help others when needed.	• Student cooperates with the team but may not actively help solve problems. • Student sometimes expresses ideas, poses relevant questions, elaborates in response to questions, and participates in group discussions. • Student provides some feedback to team members. • Student sometimes volunteers to help others.	• Student helps the team solve problems and manage conflicts. • Student makes discussions effective by clearly expressing ideas, posing questions, responding thoughtfully to team members' questions and perspectives. • Student gives useful feedback to others so they can improve their work. • Student volunteers to help others if needed.	
PROFESSIONALISM AND RESPECT FOR TEAM MEMBERS	• Student is impolite or disrespectful to other team members. • Student does not acknowledge or respect others' ideas and perspectives.	• Student is usually polite and respectful to other team members. • Student usually acknowledges and respects others' ideas and perspectives.	• Student is consistently polite and respectful to other team members. • Student consistently acknowledges and respects others' ideas and perspectives.	

Team Name: _____

Maglevacation Train Prototype Design Rubric (20 points possible)

Team Performance	Below Standard (0–1)	Approaching Standard (2–3)	Meets or Exceeds Standard (4–5)	Team Score
CREATIVITY AND INNOVATION	• Design reflects little creativity with use of materials, shows lack of understanding of project purpose, and has no innovative design features. • Design is impractical. • Design has several elements that do not fit.	• Design reflects some creativity with use of materials, shows a basic understanding of project purpose, and has limited innovative design features. • Design is limited in practicality and function. • Design has some interesting elements but may be excessive or inappropriate.	• Design reflects creative use of materials, shows a sound understanding of project purpose, and has distinct innovative design features. • Design is practical and functional. • Design is well crafted and includes interesting elements that are appropriate for the purpose.	
CONCEPTUAL UNDERSTANDING	• Design incorporates no or few features that reflect understanding of concepts (magnetism and speed).	• Design incorporates some features that reflect a limited understanding of concepts (magnetism and speed).	• Design incorporates several features that reflect a sound understanding of concepts (magnetism and speed).	
DESIGNED WITHIN SPECIFIED REQUIREMENTS	• Design violates challenge rules or specifications. • Design is not finished.	• Design meets most challenge rules and specifications. • Design is finished on time.	• Design meets all challenge rules and specifications. • Design is finished on time.	
PERFORMANCE	• Vehicle does not function or faces substantial problems in traveling the required distance.	• Vehicle functions but does not travel the required distance.	• Vehicle travels the required distance.	

TOTAL SCORE: _____

TRANSFORMING LEARNING WITH TRANSPORTATION IN THE FUTURE AND THE *STEM ROAD MAP CURRICULUM SERIES*

Carla C. Johnson

This chapter serves as a conclusion to the Transportation in the Future integrated STEM curriculum module, but it is just the beginning of the transformation of your classroom that is possible through use of the *STEM Road Map Curriculum Series*. In this book, many key resources have been provided to make learning meaningful for your students through integration of science, technology, engineering, and mathematics, as well as social studies and English language arts, into powerful problem- and project-based instruction. First, the Transportation in the Future curriculum is grounded in the latest theory of learning for children in elementary school specifically. Second, as your students work through this module, they engage in using the engineering design process (EDP) and build prototypes like engineers and STEM professionals in the real world. Third, students acquire important knowledge and skills grounded in national academic standards in mathematics, English language arts, science, and 21st century skills that will enable their learning to be deeper, retained longer, and applied throughout, illustrating the critical connections within and across disciplines. Finally, authentic formative assessments, including strategies for differentiation and addressing misconceptions, are embedded within the curriculum activities.

The Transportation in the Future curriculum in the Innovation and Progress STEM Road Map theme can be used in self-contained elementary classrooms where there is only one teacher or expanded to include multiple teachers and content areas across classrooms. Through the exploration of the Maglevacation Train Challenge, students

engage in a real-world STEM problem on the first day of instruction and gather necessary knowledge and skills along the way in the context of solving the problem.

The other topics in the *STEM Road Map Curriculum Series* are designed in a similar manner, and NSTA Press plans to publish additional volumes in this series for this and other grade levels. The tentative list of books includes the following themes and subjects:

- Innovation and Progress
  - Amusement park of the future
  - Harnessing solar energy
  - Wind energy
- The Represented World
  - Rainwater analysis
  - Recreational STEM: Swing set makeover
- Sustainable Systems
  - Hydropower efficiency
  - Composting: reduce, reuse, recycle
- Optimizing the Human Condition
  - Water conservation: Think global, act local

If you are interested in professional development opportunities focused on the STEM Road Map specifically or integrated STEM or STEM programs and schools overall, contact the lead editor of this project, Dr. Carla C. Johnson (*carlacjohnson@purdue.edu*), associate dean and professor of science education at Purdue University. Someone from the team will be in touch to design a program that will meet your individual, school, or district needs.

APPENDIX

CONTENT STANDARDS ADDRESSED IN THIS MODULE

NEXT GENERATION SCIENCE STANDARDS

Table A1 (p. 172) lists the science and engineering practices, disciplinary core ideas, and crosscutting concepts this module adresses. The supported performance expectations are as follows:

- 3-PS2-3. Ask questions to determine cause and effect relationships of electric or magnetic interactions between two objects not in contact with each other.

- 3-PS2-4. Define a simple design problem that can be solved by applying scientific ideas about magnets.

- 3-5-ETS1-1. Define a simple design problem reflecting a need or a want that includes specified criteria for success and constraints on materials, time, or cost.

- 3-5-ETS1-2. Generate and compare multiple possible solutions to a problem based on how well each is likely to meet the criteria and constraints of the problem.

- 3-5-ETS1-3. Plan and carry out fair tests in which variables are controlled and failure points are considered to identify aspects of a model or prototype that can be improved.

APPENDIX

Table A1. *Next Generation Science Standards (NGSS)*

Science and Engineering Practices

ASKING QUESTIONS AND DEFINING PROBLEMS

- Ask questions about what would happen if a variable is changed.
- Identify scientific (testable) and non-scientific (non-testable) questions.
- Ask questions that can be investigated and predict reasonable outcomes based on patterns such as cause and effect relationships.
- Use prior knowledge to describe problems that can be solved.
- Define a simple design problem that can be solved through the development of an object, tool, process, or system and includes several criteria for success and constraints on materials, time, or cost.

DEVELOPING AND USING MODELS

- Identify limitations of models.
- Collaboratively develop and/or revise a model based on evidence that shows the relationships among variables for frequent and regular occurring events.
- Develop a model using an analogy, example, or abstract representation to describe a scientific principle or design solution.
- Develop and/or use models to describe and/or predict phenomena.
- Develop a diagram or simple physical prototype to convey a proposed object, tool, or process.
- Use a model to test cause and effect relationships or interactions concerning the functioning of a natural or designed system.

PLANNING AND CARRYING OUT INVESTIGATIONS

- Plan and conduct an investigation collaboratively to produce data to serve as the basis for evidence, using fair tests in which variables are controlled and the number of trials considered.
- Evaluate appropriate methods and/or tools for collecting data.
- Make observations and/or measurements to produce data to serve as the basis for evidence for an explanation of a phenomenon or test a design solution.
- Make predictions about what would happen if a variable changes.
- Test two different models of the same proposed object, tool, or process to determine which better meets criteria for success.

ANALYZING AND INTERPRETING DATA

- Represent data in tables and/or various graphical displays (bar graphs, pictographs and/or pie charts) to reveal patterns that indicate relationships.

Table A1. (*continued*)

Science and Engineering Practices (*continued*)

ANALYZING AND INTERPRETING DATA (*continued*)

- Analyze and interpret data to make sense of phenomena, using logical reasoning, mathematics, and/or computation.

- Compare and contrast data collected by different groups in order to discuss similarities and differences in their findings.

- Analyze data to refine a problem statement or the design of a proposed object, tool, or process.

- Use data to evaluate and refine design solutions.

USING MATHEMATICAL AND COMPUTATIONAL THINKING

- Decide if qualitative or quantitative data are best to determine whether a proposed object or tool meets criteria for success.

- Organize simple data sets to reveal patterns that suggest relationships.

- Describe, measure, estimate, and/or graph quantities (e.g., area, volume, weight, time) to address scientific and engineering questions and problems.

- Create and/or use graphs and/or charts generated from simple algorithms to compare alternative solutions to an engineering problem.

CONSTRUCTING EXPLANATIONS AND DESIGNING SOLUTIONS

- Construct an explanation of observed relationships (e.g., the distribution of plants in the back yard).

- Use evidence (e.g., measurements, observations, patterns) to construct or support an explanation or design a solution to a problem.

- Identify the evidence that supports particular points in an explanation.

- Apply scientific ideas to solve design problems.

- Generate and compare multiple solutions to a problem based on how well they meet the criteria and constraints of the design solution.

ENGAGING IN ARGUMENT FROM EVIDENCE

- Compare and refine arguments based on an evaluation of the evidence presented.

- Distinguish among facts, reasoned judgment based on research findings, and speculation in an explanation.

- Respectfully provide and receive critiques from peers about a proposed procedure, explanation, or model by citing relevant evidence and posing specific questions.

- Construct and/or support an argument with evidence, data, and/or a model.

- Use data to evaluate claims about cause and effect.

- Make a claim about the merit of a solution to a problem by citing relevant evidence about how it meets the criteria and constraints of the problem.

(continued)

Table A1. (*continued*)

Science and Engineering Practices (*continued*)

OBTAINING, EVALUATING, AND COMMUNICATING INFORMATION

- Read and comprehend grade-appropriate complex texts and/or other reliable media to summarize and obtain scientific and technical ideas and describe how they are supported by evidence.

- Compare and/or combine across complex texts and/or other reliable media to support the engagement in other scientific and/or engineering practices.

- Combine information in written text with that contained in corresponding tables, diagrams, and/or charts to support the engagement in other scientific and/or engineering practices.

- Obtain and combine information from books and/or other reliable media to explain phenomena or solutions to a design problem.

- Communicate scientific and/or technical information orally and/or in written formats, including various forms of media as well as tables, diagrams, and charts.

Disciplinary Core Ideas

PS2.A: FORCES AND MOTION

- Each force acts on one particular object and has both strength and a direction. An object at rest typically has multiple forces acting on it, but they add to give zero net force on the object. Forces that do not sum to zero can cause changes in the object's speed or direction of motion.

- The patterns of an object's motion in various situations can be observed and measured; when that past motion exhibits a regular pattern, future motion can be predicted from it.

PS2.B: TYPES OF INTERACTIONS

- Objects in contact exert forces on each other.

- Electric, and magnetic forces between a pair of objects do not require that the objects be in contact. The sizes of the forces in each situation depend on the properties of the objects and their distances apart and, for forces between two magnets, on their orientation relative to each other.

ETS1.A: DEFINING AND DELIMITING ENGINEERING PROBLEMS

- Possible solutions to a problem are limited by available materials and resources (constraints). The success of a designed solution is determined by considering the desired features of a solution (criteria). Different proposals for solutions can be compared on the basis of how well each one meets the specified criteria for success or how well each takes the constraints into account. (3-5-ETS1-1)

ETS1.B: DEVELOPING POSSIBLE SOLUTIONS

- Research on a problem should be carried out before beginning to design a solution. Testing a solution involves investigating how well it performs under a range of likely conditions. (3-5-ETS1-2)

Table A1. (*continued*)

Disciplinary Core Ideas (*continued*)

ETS1.B: DEVELOPING POSSIBLE SOLUTIONS (*continued*)

- At whatever stage, communicating with peers about proposed solutions is an important part of the design process, and shared ideas can lead to improved designs. (3-5-ETS1-2)
- Tests are often designed to identify failure points or difficulties, which suggest the elements of the design that need to be improved. (3-5-ETS1-3)

ETS1.C: OPTIMIZING THE DESIGN SOLUTION

- Different solutions need to be tested in order to determine which of them best solves the problem, given the criteria and the constraints. (3-5-ETS1-3)

Crosscutting Concepts

PATTERNS

- Similarities and differences in patterns can be used to sort, classify, communicate, and analyze simple rates of change for natural phenomena and designed products.
- Patterns of change can be used to make predictions.
- Patterns can be used as evidence to support an explanation

SCALE, PROPORTION, AND QUANTITY

- Standard units are used to measure and describe physical quantities such as weight, time, temperature, and volume

SYSTEMS AND SYSTEM MODELS

- A system is a group of related parts that make up a whole and can carry out functions its individual parts cannot.
- A system can be described in terms of its components and their interactions.

ENERGY AND MATTER

- Energy can be transferred in various ways and between objects.

CAUSE AND EFFECT

- Cause-and-effect relationships are routinely identified, tested, and used to explain change.

STABILITY AND CHANGE

- Change is measured in terms of differences over time and may occur at different rates.

INFLUENCE OF SCIENCE, ENGINEERING, AND TECHNOLOGY ON SOCIETY AND THE NATURAL WORLD

- People's needs and wants change over time, as do their demands for new and improved technologies.
- Engineers improve existing technologies or develop new ones to increase their benefits, decrease known risks, and meet societal demands.

(continued)

APPENDIX

Table A2. Common Core Mathematics and English Language Arts (ELA) Standards

MATHEMATICAL PRACTICES

- MP1. Make sense of problems and persevere in solving them.
- MP2. Reason abstractly and quantitatively.
- MP4. Model with mathematics.
- MP5. Use appropriate tools strategically.
- MP6. Attend to precision.

MATHEMATICAL CONTENT

- NBT.A.2. Fluently add and subtract within 1000 using strategies and algorithms based on place value, properties of operations, and/or the relationship between addition and subtraction.
- 3.OA.A.3. Use multiplication and division within 100 to solve word problems in situations involving equal groups, arrays, and measurement quantities, e.g., by using drawings and equations with a symbol for the unknown number to represent the problem.
- 3.OA.B.5. Apply properties of operations as strategies to multiply and divide.
- 3.MD.A.1. Tell and write time to the nearest minute and measure time intervals in minutes. Solve word problems involving addition and subtraction of time intervals in minutes, e.g., by representing the problem on a number line diagram.
- 3.MD.B.4. Generate measurement data by measuring lengths using rulers marked with halves and fourths of an inch. Show the data by making a line plot, where the horizontal scale is marked off in appropriate units—whole numbers, halves, or quarters.

READING STANDARDS

- RI.3.1. Ask and answer questions to demonstrate understanding of a text, referring explicitly to the text as the basis for the answers.
- RI.3.3. Describe the relationship between a series of historical events, scientific ideas or concepts, or steps in technical procedures in a text, using language that pertains to time, sequence, and cause/effect.
- RI.3.8. Describe the logical connection between particular sentences and paragraphs in a text (e.g., comparison, cause/effect, first/second/third in a sequence).
- SL.3.6. Speak in complete sentences when appropriate to task and situation in order to provide requested detail or clarification.

WRITING STANDARDS

- W.3.1. Write opinion pieces on topics or texts, supporting a point of view with reasons.
- W.3.1.A. Introduce the topic or text they are writing about, state an opinion, and create an organizational structure that lists reasons.
- W.3.1.B. Provide reasons that support the opinion.
- W.3.1.C. Use linking words and phrases (e.g., *because, therefore, since, for example*) to connect opinion and reasons.
- W.3.2. Write informative/explanatory texts to examine a topic and convey ideas and information clearly.
- W.3.2.B. Develop the topic with facts, definitions, and details.
- W.3.3. Write narratives to develop real or imagined experiences or events using effective technique, descriptive details, and clear event sequences.
- W.3.7. Conduct short research projects that build knowledge about a topic.
- W.3.8. Recall information from experiences or gather information from print and digital sources; take brief notes on sources and sort evidence into provided categories.

SPEAKING AND LISTENING STANDARDS

- SL.3.1. Engage effectively in a range of collaborative discussions (one-on-one, in groups, and teacher-led) with diverse partners on *grade 3 topics and texts*, building on others' ideas and expressing their own clearly.
- SL3.1.D. Explain their ideas and understanding in light of the discussion.
- SL3.3. Ask and answer questions about information from a speaker, offering appropriate elaboration and detail.

PRESENTATION OF KNOWLEDGE AND IDEAS

- SL3.4. Report on a topic or text, tell a story, or recount an experience with appropriate facts and relevant, descriptive details, speaking clearly at an understandable pace.

NATIONAL SCIENCE TEACHERS ASSOCIATION

Table A3. 21st Century Skills From the Framework for 21st Century Learning

INTERDISCIPLINARY THEMES	• Inventions, History, Engineering Design Process, and Progress	• Lessons draw connections between train history and historical and current needs for progress and inventions. • Teachers highlight the importance of trains in U.S. history, particularly in regard to western expansion.	• Students interpret, organize, and present information from U.S. history in clear and complete effective formats. • Students work in collaborative teams to explore a historical topic or event and the relationship of trains to that topic or event.
LEARNING AND INNOVATION SKILLS	• Creativity and Innovation • Critical Thinking and Problem Solving • Communication and Collaboration	• Use examples and inquiry activities to connect human needs with innovation. • Teach and facilitate student use of the Engineering Design Process throughout the module. • Facilitate group work and instruct students on internet search procedures and strategies.	• Student teams implement the EDP and are able to create and present a prototype with evidence of collaboration. • Presentations reflect critical thinking and are used to draw connections between concepts and their application in group projects.
INFORMATION, MEDIA AND TECHNOLOGY SKILLS	• Information Literacy • Media Literacy • ICT Literacy	• Students use technology to conduct research on their vacation destination and a historical topic or event. • Scaffold student work in creating a media presentation.	• Student presentations include supportive evidence from research and multimedia presentation techniques.
LIFE AND CAREER SKILLS	• Flexibility and Adaptability • Initiative and Self-Direction • Social and Cross Cultural Skills • Productivity and Accountability • Leadership and Responsibility	• Scaffold prototype design work through a series of inquiry activities and topical research projects. • Use EDP to encourage flexibility (through redesign), aid time management and goal management, and structure group work. • Provide guidelines and practice opportunities for student presentations emphasizing professionalism and inclusion of all team members.	• Team projects are completed on time with evidence of participation by all team members. • Students' presentations include appropriate language and vocabulary embedded within module. • Students are able to respond to questions regarding their design process and teamwork.

Table A4. English Language Development Standards

ELD STANDARD 1: SOCIAL AND INSTRUCTIONAL LANGUAGE

English language learners communicate for Social and Instructional purposes within the school setting.

ELD STANDARD 2: THE LANGUAGE OF LANGUAGE ARTS

English language learners communicate information, ideas, and concepts necessary for academic success in the content area of Language Arts.

ELD STANDARD 3: THE LANGUAGE OF MATHEMATICS

English language learners communicate information, ideas, and concepts necessary for academic success in the content area of Mathematics.

ELD STANDARD 4: THE LANGUAGE OF SCIENCE

English language learners communicate information, ideas, and concepts necessary for academic success in the content area of Science.

ELD STANDARD 5: THE LANGUAGE OF SOCIAL STUDIES

English language learners communicate information, ideas, and concepts necessary for academic success in the content area of Social Studies.

Source: World-Class Instructional Design and Assessment Consortium (WIDA), 2012, 2012 Amplification of the English language development standards: Kindergarten–grade 12, *www.wida. us/standards/eld.aspx.*

INDEX

Page numbers printed in **boldface type** indicate tables, figures, or handouts.

INDEX

INDEX

INDEX